A Comprehensive Guide to the Biology of Cancer

A Comprehensive Guide to the Biology of Cancer:

The Crab's Legs

By

Carlos Rocha

Translated by Herald dos Reis

Cambridge Scholars Publishing

A Comprehensive Guide to the Biology of Cancer: The Crab's Legs

By Carlos Rocha

This book first published 2025

Cambridge Scholars Publishing

Lady Stephenson Library, Newcastle upon Tyne, NE6 2PA, UK

British Library Cataloguing in Publication Data
A catalogue record for this book is available from the British Library

ISBN: 978-1-0364-4831-8
ISBN (Ebook): 978-1-0364-4832-5

Contents

FOREWORD

This book evolves into an international edition, updated while maintaining the same invitation as its first version published a decade ago: to seek an understanding of how a normal cell can transform into a tumour cell. Its content introduces the reader to the complexity of cancer through the history of cell theory, retracing the series of discoveries and records that led the scientific community to develop the field of oncology. Curious phenomena, such as the occurrence of cancer in invertebrates, are provided in detail. An interesting compilation on the various types of carcinogens and their effects can also be found. The book's didactic approach is a strong point, as illustrated throughout the chapters with diagrams that facilitate the understanding of complex mechanisms, such as the transduction of extracellular signals that allow the cell to enter mitosis. In a comprehensive manner, this compendium offers readers the basic principles of cytology, which are essential for understanding the various genetic and epigenetic changes that accumulate in the cell's genome and initiate carcinogenesis. Mutations affecting genes that regulate the cell cycle, such as tumour suppressor genes and proto-oncogenes, are explained using major cancer types as models. The key abnormalities that enable cells to acquire the ten functional capabilities, necessary for tumour formation, are clearly presented through an intriguing analogy between the ten "legs" of the crab and the ten "hallmarks" of cancer (from the Latin word 'cancer', meaning crab). Although it is a book on basic oncology, some aspects of prevention, diagnosis, and treatment of malignancies are addressed. Additionally, this treatise delves into the study of natural products with antitumor activity, elucidating the mechanisms of anticancer drugs. As "the crab's legs" approach the end, we realize we have gained the knowledge to develop a broad perspective on uncontrolled cell growth.

Dr. Rommel Burbano
Coordinator of the Molecular Biology Laboratory at the Ophir Loyola Hospital.
Full Professor at the Federal University of Pará, Belém, Pará, Brazil.

1. CANCER: GENERAL CONCEPTS AND HISTORICAL SUMMARY

Conceptual introduction

Cancer is not just a disease, but the denomination applied to more than a hundred pathologies resulting from the loss of control of cell divisions in one or more parts of the body. Due to this loss of control, the mass of proliferating cells, in addition to expanding in the original tissue, takes on the characteristic of invading neighbouring tissues and, later, some distant tissues and organs.

The name cancer derives from the fact that diseased cells proliferate and infiltrate healthy cells like the legs of a crab (*cancer* in Latin, *karkinos* in Greek). Although the origin of the name cancer cannot be explained unequivocally, this relationship between the tumour and the crustacean was probably made by Hippocrates (460-370 BC), around 400 years BC.

The medical specialty dedicated to the study, diagnosis and treatment of cancers is called Oncology or Cancerology. As we begin this brief tour of Oncology, some concepts such as metaplasia, dysplasia and neoplasia are essential. Although cancer is more effectively related only to this last concept, the first two are of capital importance to avoid misunderstandings when exchanging information on the subject.

Metaplasia is the change of one type of adult tissue into another tissue of the same lineage. This process is reversible: once the cause ceases, the tissue assumes its original pattern. An interesting example, shown in Figure 1-1, occurs with the ciliated pseudostratified epithelium that lines the bronchi and which, due to smoking, transforms into stratified squamous epithelium.

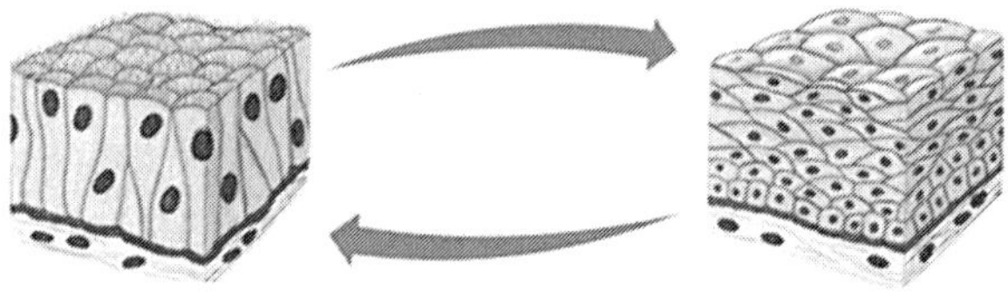

Figure 1-1. Metaplasia. Scheme of the reversible transformation of an epithelium.

It is worth remembering that the expression "adult tissue", used above, means the same as specialized or differentiated tissue. The process of cellular differentiation, which will be discussed later in a particular chapter, includes the activation and deactivation of specific genes, responsible for the characteristics that cells in a tissue must assume to become able to perform their specific functions.

Dysplasia corresponds to a transitional state between normal cell growth (completely benign) and growth considered pre-malignant. In dysplasia, the tissue is slightly abnormal, that is, the cells in general are cytologically abnormal. Among the cytological changes, increased mitotic activity and enlargement and darkening of the nucleus stand out.

Neoplasia is the localized proliferation of cells, with autonomous growth and loss of differentiation (specialization). When the proliferative process occurs in solid organs, the largest number of resulting cells forms a tumour. Many oncologists reserve the term 'neoplasia' for malignant tumours, using it as a synonym for cancer, regardless of location or symptoms.

> As Virchow examined the architecture of cancers, growth often seemed to have acquired a life of its own, as if the cells had become possessed by a new and mysterious drive to grow. This was not just ordinary growth, but growth redefined, growth almost in a new form. With some foresight [...] he called it a *neoplasia* – new, inexplicable, and distorted growth – a word that would echo throughout the history of cancer (MUKHERJEE, 2012, p. 32).

Hematopoietic tissues, that is, the tissues that form blood cells, can also be affected by neoplastic processes, generally malignant. Upon entering the circulation, leukaemia (popularized as “blood cancer”) sets in. Leukemic cells can infiltrate different organs and form secondary tumours, similar to metastases from other types of cancer. The lymphoma affects the lymphatic system.

The science of Histopathology studies how a specific disease affects a tissue, that is, it analyses tissue changes and provides information regarding the origin of the process and its relationships with normal tissue. Thanks to this science, it became possible to understand the relationships between the clinical behaviour of a tumour (the effects of the tumour on the patient) and its microscopic characteristics.

In the histopathological study of tumours, an extremely important fact is the criterion that distinguishes tumours into two general categories according to the degree of aggressiveness of their growth. Benign tumours are those that grow locally without invading adjacent tissues. Others, which invade nearby tissues and spread metastases, are malignant tumours. More details in the table below.

Table 1-1. Key differences between benign and malignant tumours.

BENIGN TUMOURS	MALIGNANT TUMOURS
Normal-appearing chromatin	**Increased chromatin**
Slow growth	**Rapid growth**
With capsule	**Without capsule**
Few vessels	**Many vessels**
No prognosis of metastasis	**With prognosis of metastasis**

Metastases are new tumour lesions from primary lesions, but without continuity with them. The term metastasis (from the Greek 'meta' meaning 'beyond' and 'stasis' meaning 'standing still'), in Latin) was used for the first time in 1829, by the French doctor Joseph Recamier (1774-1852). He was the first to describe local infiltration, vascular invasion, and development of metastases in patients with breast cancer.

For metastases to form, the initial event is the escape of some cancer cells from the primary tumour into the blood or lymphatic circulation. In addition to surviving inside the vessels, these malignant cells must be able to exit the vascular compartment, settle in another organ, and proliferate there, giving rise to the secondary form of the tumour – metastatic neoplasia.

An essential phenomenon for the growth and survival of a primary tumour and the metastases originating from it is angiogenesis, that is, the proliferation of new vessels. These newly formed vessels can either penetrate the tumour or be peripheral. Without the occurrence of angiogenesis, the tumour could not develop fully, much less promote its spread to other organs.

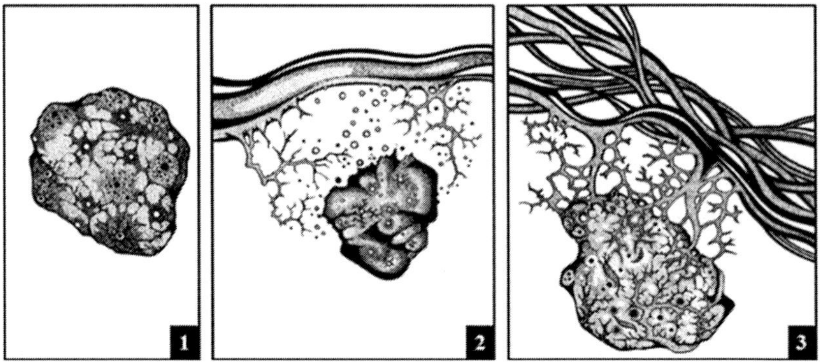

Figure 1-2. Angiogenesis. The formation of new vessels is necessary for tumour growth and metastasis. 1) Microscopic primary tumour; 2) Release of signalling molecules for the formation of capillaries; 3) Newly formed vessels, tumour growth and migration of neoplastic cells through the blood. Drawing by Daniel Ferraz (2015).

Once stimulated by chemical signals from angiogenic factors, endothelial cells proliferate to build the cylindrical walls of capillaries. This proliferation generates a capillary network with fragmented endothelial walls embedded in a tissue composed of neoplastic cells with low intercellular adhesion which facilitates cell penetration and migration through the circulatory stream.

Summary of historical records

The bones of a dinosaur, that lived 150 million years ago, show evidence of cancer. On the other hand, the oldest malignant tumour in a hominid, probably a lymphoma, was found in 1932 by Louis Leakey (1903-1972) from the remains of a jaw dating back to 4000 BC and which could be from the skeleton of a *Homo erectus* as well as an *Australopithecus*.

Human cancer must have existed since the human species existed, but the oldest record is probably found in a papyrus from Egypt from the time of Pharaoh Djozer. Currently, it is believed that this papyrus, translated in 1930, contains the teachings of Imhotep (2655-2600 BC), a great Egyptian doctor and polymath. The content referred to breast tumours and admitted the lack of resources for their cure.

Around 440 BC, the Greek historian Herodotus (484-425 BC) recorded the case of Atossa, queen of Persia, wife of Darius, who discovered a red lump in her chest, probably a inflammatory breast cancer. Darius's physicians must have tried to treat her, but without success. Finally, a slave named Democedes convinced her to allow him to excise the tumour. It is not known whether she had tumours again.

Galen (129–200 AD), a Greek physician who was considered the leading authority in the field for around a thousand years, was the first researcher to classify skin tumours into malignant and benign and considered cancer to be an incurable disease. Although he even stated that it was possible to resolve breast cancer through surgery, as long as the tumour was superficial, and all its roots were excised.

At the archaeological site of Chiribaya, located on a sandy plain in the southern tip of Peru, there is a thousand-year-old cemetery. In it, in 1990, the professor Arthur Aufderheide (1922-2013) from the University of Minnesota, a specialist in Paleopathology, performed autopsies on 140 mummified bodies. The mummy of a young woman examined had an osteosarcoma preserved in the left forearm.

In the 17th century, among other events, we can highlight some of the first medical attempts to resolve breast cancer through surgery. Wilhelm

Fabricius Hildanus (1560-1634), the "father of German surgery," removed cancerous lymph nodes from the breast. Another German surgeon, Johannes Scultetus (1595-1645), described and performed mastectomies – breast removal surgeries.

The British doctor John Hill (1716-1775) published in 1761 a booklet in which he incriminated the excessive use of snuff (powdered tobacco used for sniffing) as being a cause of nasal mucosa cancer. Percivall Pott (1714-1788), another Briton, likely discoverer of the carcinogens, reported in 1775 the highest incidence of skin cancer in the scrotum of men who, as boys, worked as chimney sweeps.

Still in the 18th century, the experiments conducted by the Frenchmen Jean Astruc (1684-1766) and Bernard Peyrilhe (1737-1804) represented the beginnings of experimental oncology. Peyrilhe attempted the first experimental cancer transmission by injecting fluid from a human breast lesion it into the peritoneal cavity of a dog, which developed a non-cancer abscess.

In 1855, Rudolf Virchow (1821-1902) created a cellular theory of human biology, based on two fundamental principles. The first, borrowed from the idea of Schleiden (1804-1881) and Schwann (1810-1882), from 1839: like animals and plants, the human body is made of cells; the second proposed that all cells come from other cells: *omnis cellula et cellula*, in his words.

Considering that cells necessarily originate from other cells, the growth of a tissue could only occur by increasing its size or increasing its number. Virchow then called these increases hypertrophy and hyperplasia, respectively. It can be observed that, in fact, the muscles in general grow through hypertrophy, while the blood and dermis grow through hyperplasia.

This explanation was compelling and led to a new understanding of both normal and pathological growth. From Virchow onward, scientists increasingly focused on studying cells to understand cancer and its treatment. Cancer was, then, a disease of pathological hyperplasia, in which cells acquired an "autonomous desire" to divide.

By observing the pathological behaviour of leukemic cells under a microscope, Rudolf Virchow discovered the uncontrollable growth of these cells. According to Mukherjee (2012), Virchow Virchow did not coin the term, although he did create a comprehensible description of neoplasia, a very appropriate word to refer to new, distorted cell growth – cancerous growth.

Theodor Boveri (1862-1915) suggested that malignant tumours could result from chromosomal abnormalities. He arrived at the hypothesis by studying sea urchin embryos and observing that animals with highly unstable karyotypes did not survive, whereas certain chromosomal gains or losses can give a cell and its descendants the ability to multiply and give rise to cancer.

From the second half of the 19th century, oncological surgeries (and surgeries in general) became more viable, mainly due to the development of asepsis and antisepsis techniques introduced by Joseph Lister (1827-1912) in the 1860s and the acceptance of Louis Pasteur's (1822-1895) proposal on the microbial origin of infectious diseases.

Different surgical initiatives began to be successful. Examples of this advancement include the first successful removal of a stomach tumour performed 1881 by the German surgeon Theodor Billroth (1829-1894) and an extended mastectomy performed in 1890 by the American surgeon William Stewart Halsted (1852-1922) at Johns Hopkins Hospital in Baltimore, Maryland.

In 1900, Austrian gynaecologist Ernst Wertheim (1864-1920) published a work on uterine cancer focusing on a hysterectomy technique (surgery to remove the uterus) including removal of the uterus, parametrium, tissues around the upper part of the vagina and pelvic lymph nodes, which had reduced the death rate resulting from this surgery at Queen Elizabeth Hospital from 72% to 10%.

French scientist Henry Becquerel (1852-1908) was one of the collaborators in the discovery of radioactivity. He dedicated himself to the study of uranium radiation emitted in photographic films. He shared the 1903 Nobel Prize in Physics with Marie and Pierre Curie, who discovered the radioactivity of thorium and isolated two elements (polonium and radium) more active than uranium.

Although it is still a subject of debate and discussion, the use of radiotherapy must have started in 1895/96, in the United States. The first attempt aimed to treat a local recurrence of breast carcinoma (1 hour of exposure for 18 days) and was carried out by the American Emil Grubbé (1875-1960), in Chicago. The patient showed some improvement but eventually died from metastasis.

By the end of 1896, several researchers had noticed the palliative effects of rays on pain caused by some types of cancer. The medicinal application of radioactive isotopes evolved throughout the 20th century, marked by three major scientific achievements: the discovery of X-rays, the discovery of natural radioactivity and the production of artificial radioactive elements.

Marie Curie (1867-1934) received the Nobel Prize twice. In addition to the Nobel Prize in Physics in 1903, she was awarded the Nobel Prize in Chemistry in 1911 for her research into the properties of the element radium and its compounds. Her daughter Irène Curie (1897-1956) and the husband, Frédéric Joliot (1900-1958), won the Nobel Prize in Chemistry in 1935 for their discovery of artificial radioactivity.

The first antineoplastic chemotherapy drug was developed in the 1940s, from mustard gas, used in both World Wars as a chemical weapon. After exposing soldiers to this agent, they were observed to develop bone marrow and lymphoid hypoplasia. These observations have led to its use in the treatment of diseases that progress with an increase in white blood cells, such as many leukaemias and some lymphomas.

Still in the 1940s, Sidney Farber (1903-1973) used antifolates, such as aminopterin, to combat acute lymphoblastic leukaemia. Although temporarily, the remission of symptoms was achieved. His team treated 16 children, of which ten reacted. Some lived for four or six months after diagnosis. For leukaemia, at the time, six months of survival was an eternity.

Following the publications of clinical studies carried out with mustard gas (1946) and Farber's observations on the effects of antifolates in children with leukaemia (1948), there was increasing progress in oncological chemotherapy. Both works contributed greatly to showing the possibility of fighting cancer with medicines, which until then was completely inconceivable.

In 1951, George Otto Gey (1899-1979) managed to cultivate cells from uterine tumour tissue, establishing "HeLa", the first immortal human cell line, still used throughout the world in research such as the nature of various types of cancer and the development of vaccines. The fact that human tumours could give rise to lineage-continuous cells increased interest in tissue culture.

In the 1970s, the *Src* gene was discovered, the first oncogene (a gene that produces a protein capable of causing cancer). In the following years, many other oncogenes and tumour suppressor genes (which perform the opposite role, that is, stop cancer) appeared. Since the 1990s, molecular biology has been the area that has brought the most discoveries to the understanding of cancer.

Around 1990, Dennis Slamon's team developed the first targeted therapy, using Herceptin (trastuzumab), an antibody-drug conjugate targeting HER-2. Therefore, Herceptin proved to be useful in cases of breast cancer in which the HER-2 protein is overexpressed (around 30% of cases). In 1998, the drug received FDA approval for use in HER2-positive metastatic breast cancer.

In 1995, the International Agency for Research on Cancer (IARC) characterized two types of human papillomavirus (HPV), 16 and 18, as being oncogenic. HPV became official as a cause of cervical cancer and, from 1996, pharmaceutical companies began to work on developing vaccines. Almost ten years later, the results were published, and these vaccines were shown to be effective.

Reading Suggestions

The contribution of women to radiobiology: Marie Curie and beyond
Reports of Practical Oncology & Radiotherapy
Anna Gasinska

Temporary Remissions in Acute Leukemia in Children Produced by Folic Acid Antagonist, 4-Aminopteroyl-Glutamic Acid (Aminopterin)
The New England Journal of Medicine
Sidney Farbe, Louis K. Diamond, Robert D. Mercer, Robert F. Sylvester Jr., and James A. Wolff.

Film Suggestion

Living Proof

Based on real events, the film tells the story of Dr. Dennis Slamon (Harry Connick Jr.). He works on developing a drug called Herceptin®, a possible treatment in the fight against breast cancer. There are financing difficulties for his project, and philanthropists Lilly Tartikoff and Ron Perelman help make the most important work of the doctor's life a reality. This impactful story shows us the difficult fight against cancer.

2. Cancer in Non-Human Animals

Introduction

The occurrence of malignancies is not unique to human tissues, nor is it restricted to mammals. It likely affect the various animal groups. When we consider that the occurrence of neoplasms in a particular tissue results from the loss of cell cycle control, we can imagine that such incidents can occur in any living being with tissues, even some cases have not yet been described in the scientific literature.

You, who are now reading this text, may have never heard of cancer in birds, reptiles, fish, or invertebrates. However, tumour formation is a phenomenon that affects nearly all known multicellular organisms, including invertebrates and plants. The number of studies focusing on non-human neoplasms is no longer small, and some will be mentioned in this chapter.

Carcinogenesis always begins with DNA damage. Generally, this damage is exacerbated by chemical, physical, or biological agents. A major concern today regarding the causes of cancer is environmental carcinogenic factors. Among these are chemical products that expose different populations to the risk of mutations, which can eventually lead to cancer.

Cancer in invertebrates

Macoma balthica is a species of bivalve mollusk, that is, an animal belonging to the same class as oysters, mussels, clams, and scallops. This small marine mollusk is commonly known as the Baltic macoma due to its wide distribution in the Baltic Sea. Analysis of the chromosomes in its cells typically reveals a karyotype with 2n = 38 chromosomes.

In 2001, Catherine Thiriot-Quiévreux and Maciej Wolowicz conducted a study that assessed the prevalence of a neoplasm in the gills of *M. balthica* at various locations in the Gulf of Gdansk, in the Baltic Sea. The occurrence of this neoplasm was identified by abnormal metaphases, with a higher-than-normal chromosome count. Among the animals studied, 33% had neoplasms in their gills.

Numerous examples of neoplasms in invertebrates are archived in the Registry of Tumours in Lower Animals (RTLA) in Sterling, Virginia, (USA). Among the specimens of crustaceans, for example, notable cases include carcinomas of the antennal gland and of posterior intestine in crabs. In contrast, the occurrence of malignant tumours in most lobsters is very rare.

Early in the 20th century, the German embryologist Theodor Boveri correlated chromosomal abnormalities, known as aneuploidies, with cancer. He studied sea urchin embryos and observed that even with a highly unstable karyotype, they survived. However, in some cases, the gain or loss of chromosomes could give a cell and its progeny the ability to proliferate and form tumours.

Cancer in vertebrates

Neoplasms in fish, like those in higher vertebrates, can arise in all organs, tissues, and cell types. Skin tumours, especially papillomas, are the most frequently observed in fish. Furthermore, the study of tumours in fish is of great importance in terms of comparative oncology, as they can serve as models to help understand similar conditions in humans.

Another aspect to consider, that supports the study of neoplasms in fish, is their possible relation to pollution and the fact that they can be used as indicators of the existence of potentially dangerous environmental carcinogens for higher vertebrates, including humans. One must also not forget the difficulty or even impossibility of commercializing fish with visible tumours.

The description of spontaneous neoplasms in fish is not a recent activity; it has been carried out for several decades. An interesting example from the early 1980s is the work of Meyers and Hendricks, who presented histopathological descriptions of four neoplasms (lymphoma, cutaneous fibrosarcoma, dermal haemangioma, and renal cystadenoma) in three species of the family Salmonidae in North America.

The fish shown in the following figure is a breeding sea bass (*Dicentrarchus labrax*) kept under experimental conditions in in Lisbon (Portugal). A neoplastic mass can be seen at the upper commissure of the right operculum. Histopathological examination allowed for the diagnosis of a cancer called cutaneous fibropapilloma, with no evidence of metastasis.

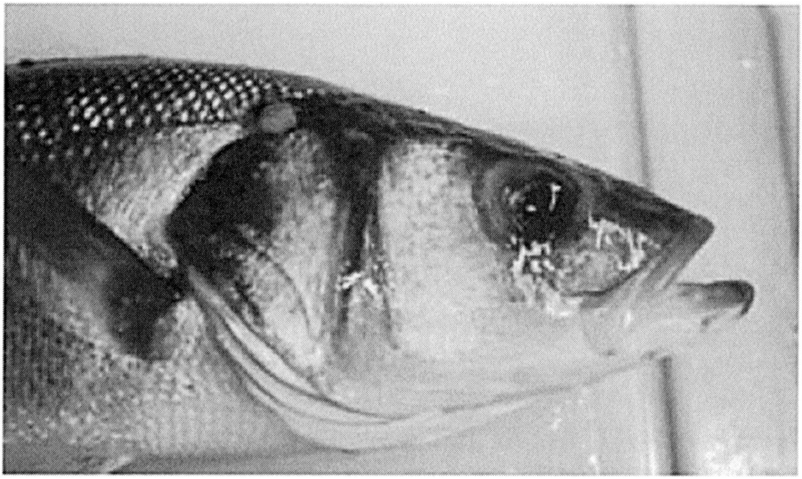

Figure 2-1. Cancer in fish. Specimen of *D. labrax* with nodular formation above the right operculum. Source: Ramos and Peleteiro (2003).

This case of cancer in sea bass was reported by Paula Ramos and Maria da Conceição Peleteiro in a 2003 publication in the *Revista Portuguesa de Ciências Veterinárias*. In the same article, two other cases of spontaneous neoplasms in fish were presented: leiomyoma originating from the smooth muscle tissue of the intestinal wall of sardines; and multiple melanomas of the skin in an aquarium carp.

In medical research, a small tropical freshwater fish of the species *Danio rerio* (zebrafish) has gained attention in recent decades. The biologist

George Streisinger from the University of Oregon (USA) is credited with introducing of zebrafish as an animal model in scientific research. Regarding cancer, lymphoid leukaemia, melanoma, and hepatocarcinoma have already been described in zebrafish, demonstrating that the molecular mechanisms involved are similar to those in humans.

In the leopard frog (*Rana pipiens*), a North American native amphibian, there is an example of neoplasia whose development depends on the presence of oncoviruses (viruses capable of inducing tumour formation). Renal adenocarcinoma (or Lucke's tumour) occurs frequently in wild populations of leopard frogs, with evidence suggesting that a herpesvirus plays a significant role in the aetiology of this tumour.

Kingsnakes, such as those of the species Lampropeltis getula, are interesting snakes of North American origin, often kept as a pet and is so named because of its habit of preying on other snakes in the wild, though it also consumes eggs, birds, and rodents. A cutaneous mastocytoma, a tumour characterized by an abnormal accumulation of mast cells in the skin tissues, has been found in individuals of this species.

Mast cells are connective tissue cells that play a crucial role in allergic reactions, and they also act against blood clot formation and pathogens. This neoplasm in the kingsnake was described in 1998 by Juergen Schumacher and colleagues from the College of Veterinary Medicine at the University of Florida (USA). The patient was a 16-year-old captive male.

Fibropapillomatosis is a disease that affects sea turtles, predominantly the green sea turtle (*Chelonia mydas*), found in both the Atlantic and Pacific Oceans. The predominant lesions associated with this condition are fibromas, cutaneous papillomas, and fibropapillomas. Cutaneous tumours mainly appear at the base of the limbs, tail, neck, and head, including the eyes.

The exact aetiology of fibropapillomatosis has not yet been established, but it is believed to result from a viral infection and that its manifestation is related to certain predisposing factors such as genetics, environmental changes, injuries, and stress. Recently, the use of polymerase chain reaction (PCR) in neoplastic tissue has detected the presence of herpesvirus DNA.

Parakeets, parrots, and macaws (family Psittacidae) have become popular pets. In 2020, Morgan Halley and colleagues from Georgia (USA) reported the occurrence of squamous cell carcinoma (SCC) in two Psittacidae birds: a green-winged macaw (*Ara chloropterus*) and a black-headed parrot (*Pionites melanocephalus*). In both cases, the tumour affected the beak and mandibular bone.

In mammals, SCC is a malignant tumour that can develop from keratinocytes, the skin cells responsible for producing keratin. In 2021, Aycan Özcan-Martz and colleagues from the Clinic for Ruminants and Swine in Leipzig (Germany) presented a study including six sick cattle. Two cows with ocular SCC were euthanized. The prognosis of SCC depends on early clinical detection, the presence of metastases, and timely tumour removal.

Also in 2021, in Japan, Yasutsugu Miwa and colleagues completed a retrospective study (2005 to 2015) of a series of spontaneous oral tumours in 18 rabbits, whose most common clinical sign was feeding anomalies. Thirteen different types of tumours were diagnosed, with squamous cell carcinoma, fibrosarcoma, and osteosarcoma being the most frequent. The study also found that surgical excision and radiotherapy were effective in rabbits.

Mastocytomas, like the tumour mentioned in the kingsnake quoted above, are cutaneous conditions characterized by a single isolated lesion, usually nodular, occasionally in plaque form, with a pinkish-yellow or reddish-brown appearance. The lesions present itching, oedema with erythema, and blister formation. These tumours occur in various mammals and are the most frequent neoplasms in dogs.

In 2018, Ana Carolina Souza and colleagues from the Veterinary Hospital of the School of Agrarian and Veterinary Sciences of UNESP, in Jaboticabal (Brazil), presented a retrospective study of cases of cutaneous mastocytomas in 192 dogs treated at that hospital between 2005 and 2015. The highest incidence was observed in mixed-breed dogs, followed by boxers. The tumours were more frequent in the limbs, and each patient usually presented with more than one nodule.

Canine mastocytomas rarely arise from the oral mucosa. However, in these cases, they show a more aggressive clinical course compared to those occurring in the skin. In 2013, James W. Elliott from Veterinary Clinical Science at Liverpool University (England) and other researchers from the United Kingdom, conducted a retrospective study of 33 dogs with mastocytomas originating from the oral mucosa. They found a a high incidence of lymph node metastasis at diagnosis, contributing to a poor prognosis.

Carcinomas are neoplasms that originate from epithelial tissue, and, in the stomach, they are the most frequent type of cancer. However, reports of gastric carcinoma in dogs are relatively rare, although the risk is elevated in certain breeds, such as the Chinese Chow Chow and the Belgian Tervuren Shepherd. The Chow Chow is the famous dog with a blue tongue and a mane resembling that of a lion. This Chinese breed reached the West through trade between China and England.

Amy M. Koterbay and colleagues from the University College Dublin, Veterinary Hospital, in Dublin (Ireland), reviewed the Veterinary Medical Database (VMDB) in 2020 and analysed 106 Chow Chow dogs with gastric carcinoma. Most of affected dogs were euthanized, without treatment, within two weeks of diagnosis. However, some dogs survived significantly longer – up to 36 months – when treated with surgery and chemotherapy.

Lymphoma is a malignant hematopoietic neoplasm that originates from lymphoid cells in solid organs such as the lymph nodes, liver, and intestines. In cats (*Felis catus*), reported risk factors for developing lymphoma include infections with feline leukaemia virus and/or feline immunodeficiency virus, as well as exposure to tobacco smoke, radon, agricultural chemicals, and immunosuppressive treatments.

Regarded as the most common neoplasm in domestic cats, lymphoma typically develops in cats with an average age of 11 years. In 2012, Peter F. Moore and colleagues from the School of Veterinary Medicine,

University of California (USA), published a study identifying 120 cats with gastrointestinal lymphomas between 1995 and 2006. The lymphomas were classified according to the World Health Organization (WHO) criteria, revealing a predominance (n = 84) of mucosal T-cell lymphoma.

A survey conducted in 2012 at the São Paulo Zoo (Brazil) revealed that over a period of 10 years, there were 30 cases of cancer in mammals, seven in birds, and six in reptiles. As non-human animals are also victims of neoplasms, the Zoo, in partnership with the A.C. Camargo Hospital, initiated a project to establish a tumour bank for the animals at the institution. The biobank, a pioneering infrastructure in Latin America for the study of neoplasms in wild animals, already contains over 300 tumour samples.

According to the opinions of veterinarians and biologists, the high incidence of neoplasms in animals kept in zoos worldwide is primarily related to the aging of these animals. Since their food supply is ensured and they are free from predators and major climatic adversities, it is estimated that animals in captivity are estimated to live approximately 20% longer than their wild counterparts.

Much of what we know today about cancer has been obtained from studies using cancer cell lines and animal models. Before its application in cancer research, *in vitro* cell cultivation began in the early 20th century, pioneered by American scientist Ross Granville Harrison in 1907 and French researcher Alexis Carrel in 1912. This technique allowed for the study of animal cells outside the organism in a controlled environment.

Various mammals are used to induce of cancer through the inoculation of human neoplastic cell lines, as well as for testing carcinogenic, anticarcinogenic, and immunomodulatory drugs. A notable example is a study on experimental gastric carcinogenesis in the *Cebus apella* monkey, initiated in 2011 by Joana Costa and colleagues from the Human Cytogenetics Laboratory and João de Barros Barreto University Hospital at the Federal University of Pará, in Belém, Brazil.

That research was conducted in collaboration with the National Primate Centre of the Ministry of Health, the Federal University of Ceará, the

Federal University of São Paulo, and the Federal University of Lavras. The aim was to establish a model of gastric carcinogenesis in a non-human primate. In some monkeys, stomach tumours were induced by inoculation with cancerous cells from the human gastric cell line ACP03, while in others, tumours were induced through treatment with the carcinogen N-methyl-N-nitrosourea (MNU).

Throughout the research period (approximately 2.5 years), serum biochemistry values and haematological parameters were evaluated, along with the expression and copy number of the MYC oncogene. All tumour biopsies from animals inoculated with ACP03 showed increased production of the MYC protein. Additionally, the homeopathic immunomodulatory compound Canova was also tested. Its potential to enhance immunity may help mitigate some of the adverse effects associated with cancer treatment.

READING SUGGESTIONS

Zebrafish cancer: the state of the art and the path forward
Nature Reviews Cancer
Richard White, Kristin Rose, and Leonard Zon

Putting zoo animal cancer into perspective
Zoo Biology
Marcus Clauss and Dennis W H Müller

Animal Models of Colorectal Cancer: From Spontaneous to Genetically Engineered Models and Their Applications
Veterinary Science
Elisabete Nascimento-Gonçalves, Bruno A.L. Mendes, Rita Silva-Reis, Ana I. Faustino-Rocha, Adelina Gama, and Paula A. Oliveira.

3. CARCINOGENS

Introduction

Most tumours are sporadic, meaning they arise from *de novo* somatic genetic alterations (new mutations) caused by exposure to chemical, physical, and biological carcinogens. While chemical and physical agents are more commonly discussed, according to David Schottenfeld and Jennifer Beebe-Dimmer from the United States, biological carcinogens, such as infectious agents, are responsible for approximately 25% of all neoplasms.

Oxidative stress or chronic inflammation related to substances, physical agents, or infectious diseases leads to the initiation of continuous regenerative processes aimed at restoring the original integrity of tissues, which relies on cooperation among large groups of cells and the exchange of information between cells, much of which is mediated by growth factors.

The replacement of destroyed or damaged cells occurs through the stimulation of cell division, which creates an opportunity for the accumulation of genetic damage (accelerating the cell cycle increases the likelihood of errors in DNA replication and distribution). The emergence of dysplastic cells, characteristic of progression toward invasive tumours, is followed by the development of pre-malignant lesions, and eventually, as a result of this process, cancer appears.

Various types of cancer tend to become more prevalent in populations exposed to risk factors, including smoking, unhealthy dietary habits, exposure to chemical carcinogens, physical carcinogens in the workplace or environment, as well as infections by certain viruses and bacteria (Table 1). In this context, epidemiological studies highlight smoking as the primary carcinogenic agent.

Table 3-1. Carcinogens and Their Relationships with Cancer Types.

Chemical	Smoking	Lung cancer, upper airway cancer, oesophageal cancer, bladder cancer, and pancreatic cancer.	
	Alcohol consumption	Digestive tract cancer, respiratory tract cancer, and liver cancer.	
	High intake of saturated fats and low fibre intake	Colon cancer.	
	Metals such as Al, As, Be, Cr, Ni and Rn	Lung cancer.	
Physical	Ionizing radiation	Thyroid cancer.	
	Ultraviolet radiation	Skin cancer.	
Biological	Viruses	RSV	Rous Sarcoma (in chickens).
		HPV	Cervical cancer.
		HTLV-1	T-cell leukaemia.
	Bacteria	*Helicobacter pylori*	Gastric cancer.
		Salmonella typhi	Hepatobiliary carcinoma.

It is estimated that quitting smoking could prevent about one-third of cancer deaths worldwide. Smoking, this major villain with its polycyclic aromatic hydrocarbons and nitrosamines, is not only a risk factor for cancers of the lung, upper airways, oesophagus, bladder, and pancreas, but is also likely linked to cancers of the stomach, liver, colon, kidney, and myeloid leukaemia.

Excessive alcohol consumption is also implicated in the development of certain neoplasms. Conversely, the evidence of carcinogenicity of certain heavy and transition metals to humans has been thoroughly evaluated by the International Agency for Research on Cancer (IARC). Metals such as arsenic (As), hexavalent chromium (Cr), nickel (Ni), cadmium (Cd), lead (Pb), and their compounds are associated with cancer in humans.

The IARC is an agency of the World Health Organization (WHO) with the mission to coordinate and conduct research on the causes of cancer in humans and to develop scientific strategies for its prevention and control. The IARC classifies agents to which humans may be exposed based on evidence of their potential to pose a cancer risk to humans (Table 2).

Table 3-2. IARC Classification.

Group	Classification	Description
1	Carcinogenic to humans	Agents whose carcinogenic effect on humans has been demonstrated through cause-effect epidemiological studies.
2A	Probably carcinogenic to humans	Agents with limited evidence of carcinogenic effect on humans and sufficient evidence of carcinogenicity in animal experiments.
2B	Possibly carcinogenic to humans	Agents with limited evidence of carcinogenic effect on humans and insufficient evidence of carcinogenicity in animal experiments.
3	Not classifiable as to carcinogenicity to humans	Agents for which the evidence of carcinogenicity is inadequate in both humans and animal experiments.

Chemical carcinogens

Among the chemical agents classified in IARC Group 1, we find components of tobacco smoke (such as benzopyrene), alcoholic beverages, aflatoxins, formaldehyde, benzene, cyclophosphamide, asbestos, and metals like hexavalent chromium, cadmium, and beryllium. In Group 2A, notable agents include vinyl bromide, glyphosate, and inorganic lead compounds. In Group 2B, some examples include methylmercury, phenobarbital, and acetaldehyde.

Tobacco smoke contains multiple mutagenic agents, whereas ethanol contains little to no mutagenic power. Ethanol contributes to cancer development through its cytotoxic effects on epithelial cells, leading to constant cellular renewal. These proliferating cells might harbour changes caused by mutagens, such as those found in tobacco, which could drive them toward malignant proliferation.

Even though ethanol is not a direct-acting carcinogen, its metabolite acetaldehyde (acetic aldehyde) can promote the formation of neoplasms. Chronic alcoholics have a tenfold higher risk of developing oral cancer compared to non-drinkers. The combined abuse of ethanol and tobacco can increase the risk of developing various head and neck tumours by up to 100 times.

Aflatoxins are a type of mycotoxin, which are secondary metabolites produced by certain strains of fungi from the *Aspergillus* genus. These fungi naturally grow on food products such as peanuts, corn, beans, rice, and wheat, among others. Aflatoxin B_1 (AFB_1) exerts its toxic effects after being converted in the liver into AFB1-epoxide, which then reacts with cellular macromolecules, including proteins, RNA, and DNA.

Frequent ingestion of aflatoxins, or their consumption in large quantities, can lead to liver cancer due to their mutagenic effect on the *TP53* gene in hepatocytes. *TP53* is a tumour suppressor gene, and mutations induced by AFB_1-epoxide, which impair the production of its protein product (p53 protein), promote increased cell proliferation and tumour growth.

Chromium is a transition metal with three main oxidation states: Cr^{+2}, Cr^{+3}, and Cr^{+6}, forming various coloured compounds. Hexavalent chromium is the least stable and most biologically reactive form, known for being highly toxic, genotoxic, and carcinogenic. The film *"Erin Brockovich"*, which earned Julia Roberts an Oscar, is based on a true story: the tragic effects of chromium contamination on the population of a North American city.

The character Erin Brockovich uncovers that a company had been contaminating the town's water, leading to many residents developing cancer. As a result, the company depicted in the film was ordered to pay millions in compensation. In reality, industries are the primary culprits behind chromium pollution, and this issue is more widespread than many realize. The amount of chromium released into the environment daily is a significant concern.

Mercury, like some other metals, and certain organometallic compounds, has exhibits mutagenic properties by acting on tubulin, the structural subunit of microtubules in the spindle fibres of cell division. This action interferes with the polymerization of tubulin, disrupting the microtubule network, which leads to uncontrolled chromosome distribution. Additionally, it generates reactive oxygen species (ROS) that directly damage genetic material.

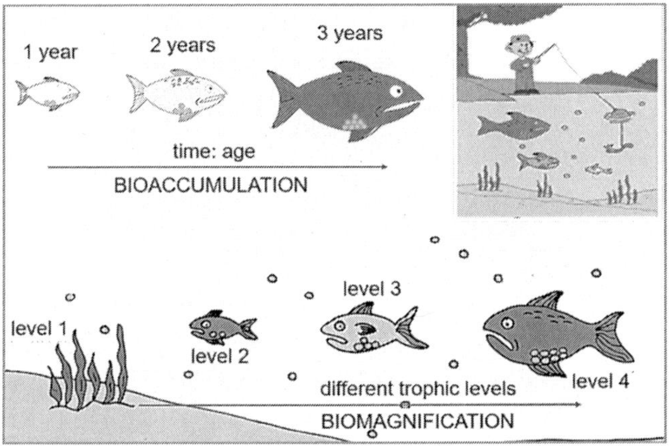

Figure 3-1. Bioaccumulation and Biomagnification. A metal or non-biodegradable chemical accumulates over the lifespan of an organism and throughout the food chain. Source: Adapted from Rocha (2009).

The liposolubility of organic mercury compounds, such as methylmercury, allows them to cross biological membranes, contributing to bioaccumulation and biomagnification along food chains. As a result, the highest concentrations of the metal are found at the upper trophic levels (the top of the food chain). The consumption of aquatic organisms contaminated with methylmercury is the primary route of human exposure.

Physical carcinogens

The electromagnetic spectrum refers to the entire range of electromagnetic radiation, divided into seven types of waves. In order of increasing wavelength, these are: gamma rays, X-rays, ultraviolet, visible light, infrared, microwaves, and radio waves. In terms of carcinogenesis, ionizing radiation (X-rays and gamma rays) is to thyroid cancer as ultraviolet radiation is to skin cancer.

Based on frequency and wavelength, the spectrum can be divided into several spectral zones. Ultraviolet rays have wavelengths (100-400 nm) shorter than those in the visible region of the human eye (400-700 nm). In fact, ultraviolet (UV) radiation is divided into three wavelength ranges: UVC (100-280 nm), UVB (280-320 nm), and UVA (320-400 nm). UVC has the peak absorption by pure DNA (260 nm), despite its limited penetration into the epidermis.

UVB is highly effective at causing direct DNA damage, photoimmunosuppression, erythema, thickening of the stratum corneum, and melanogenesis. Although UVA is less efficient at producing erythema and subsequent melanogenesis, it induces oxidative processes by reacting with molecular oxygen, producing reactive species capable of triggering inflammatory reactions in the skin and causing DNA damage. The UVA-2 range (320 to 340 nm) is more erythemogenic than UVA-1 (340 to 400 nm).

UV radiation, particularly at wavelengths around 260 nm, is strongly absorbed by DNA bases and can cause a photochemical fusion of adjacent pyrimidine bases, leading to the formation of thymine dimers. These dimers are considered mutagenic photoproducts because they can interfere with the replication process. The DNA in sun-exposed skin may acquire hundreds of these dimers each day, which are normally repaired.

DNA damaged by thymine dimer formation is repaired through a mechanism known as nucleotide excision. Although there are enzymes responsible for correcting DNA replication errors and repairing damage caused by various factors, this process is not always successful. Consequently, failures in the enzymatic repair that removes these thymine dimers can result in abnormal cell death and an increased risk of skin cancer.

The significant increase in skin cancer incidence over the past few decades has made the widespread use of sunscreen crucial for preventing skin diseases. In addition to protecting against cancer, sunscreens also help prevent photoaging, especially for those who are exposed to ultraviolet radiation for extended periods, such as outdoor workers. While the risk from extensive sun exposure is higher for people with fair skin, sunscreen is important for everyone.

Biological carcinogens

Certain infectious agents can cause malignancies or increase the risk of developing them. Some viruses, for instance, can disrupt the signalling pathways that usually regulate cell growth and proliferation. Additionally, some infections weaken the immune system, making it less effective at combating other infections and eliminating altered cells. Furthermore, certain viruses, bacteria, and parasites can cause chronic inflammation, which may lead to neoplasms.

Some of the earliest biological agents implicated in a connection between infection and carcinogenesis are two retroviruses from birds: the avian myelocytomatosis virus and the Rous sarcoma virus. The pathologist Francis Peyton Rous (1879-1970), honoured in the name of this sarcoma, demonstrated that cancer could be induced in healthy chickens infected by the virus, a discovery that earned him the Nobel Prize in Physiology or Medicine in 1966.

In humans, more than 100 different types of human papillomavirus (HPV) can infect the skin and mucous membranes, potentially forming warts (technically called condyloma acuminata). HPV infection is very common and primarily occurs in the cervix, but it can also manifest in the vagina, vulva, anus, penis, mouth, and oropharynx. In most cases, the infection is transient and resolves spontaneously.

In a small number of cases, HPV infection persists, potentially leading to the development of precancerous lesions. If these lesions are not identified and treated properly, they may progress to malignancy. Whether or not the infection progresses to cancer depends on several factors, including the specific type(s) of HPV present, with at least 13 types known to be of high oncogenic risk.

Cervical cancer is the second most common type of cancer among women living in less developed regions of the world, surpassed only by breast cancer. Certain types of HPV, such as types 16, 18, 31, 33, 35, 39, 45, 51, 52, 56, 58, 59, and 66, have already accumulated substantial evidence of carcinogenic potential. The most prevalent viral DNA types found in women with cervical carcinoma are types 16 and 18.

HTLV stands for Human T-cell leukaemia virus. There are four types, with HTLV-1 being the most associated with disease, particularly adult T-cell leukaemia/lymphoma – a typically fatal neoplasm of mature T lymphocytes that is resistant to chemotherapy. The term "leukaemias" refers to malignant derivatives of white blood cells, while "lymphomas" include solid lymphoid tumours, often found in the lymph nodes.

The term "primary liver cancer" refers to tumours that originate in the liver itself: hepatocarcinoma or hepatocellular carcinoma (accounting for over 80% of cases), cholangiocarcinoma, hepatic angiosarcoma, and hepatoblastoma. Chronic infections with hepatitis B and C viruses (HBV and HCV) are associated with most hepatocellular carcinoma cases. "Secondary liver cancer" originates in another organ and subsequently metastasizes to the liver.

Hepatocellular carcinoma originates from the liver's main cells – the hepatocytes. Cholangiocarcinoma affects the bile ducts within the liver. Hepatic angiosarcomas are tumours of the liver's blood vessels. Hepatoblastoma is a rare tumour that typically occurs in children three years old or younger. The secondary type most commonly arises from a malignant tumour in the large intestine.

Another virus linked to various benign and malignant proliferative disorders is Human Herpesvirus 4 (HHV-4), also known as Epstein-Barr Virus (EBV). EBV is commonly associated with lymphoid-related conditions, such as infectious mononucleosis and post-transplant lymphoproliferative disorder. Its oncogenic role has been extensively studied in Burkitt's lymphoma and Hodgkin's disease.

The first epithelial neoplasm to be associated with EBV was nasopharyngeal carcinoma. More recently, the involvement of EBV in the development of gastric carcinoma has also been demonstrated. The indication that EBV is linked to the oncogenic processes of both gastric and nasopharyngeal carcinomas is supported by the significant percentage of association with these tumours, with over 80% of malignancies attributed to EBV.

Carcinomas arising from epithelial cells that line cavities are known as squamous cell carcinoma (SCC), while those that develop from secretory epithelial cells are called adenocarcinomas. Adenocarcinomas account for approximately 95% of malignant tumours of the stomach and the gastric adenocarcinoma is among the most common neoplasms in the gastrointestinal tract, regardless of gender or specific regional variations.

Cancer has a multifactorial origin, and its development depends on the interaction between various genes and the environment. In this context, dietary factors can significantly increase the incidence of gastric neoplasia in certain regions, particularly high consumption of salt-preserved and smoked foods, limited use of refrigeration, and low intake of fresh fruits and vegetables.

In various types of gastrointestinal tract cancers, bacterial and viral infections that induce chronic inflammation also play a significant role. Many of these neoplasms begin with an inflammatory process and are preceded by prolonged precancerous conditions, which develop through multiple sequential steps, typically including dysplasia, a premalignant lesion, and carcinoma.

Gastric cancer primarily affects patients with long-standing atrophic gastritis, most commonly caused by *Helicobacter pylori* infection (a Group 1 carcinogen according to IARC). However, despite the high global prevalence of *H. pylori* infection, less than 3% of those infected develop gastric neoplasms. This suggests that the progression to malignancy is likely associated with bacterial, environmental, and host-related factors.

To better understand the influence of the gastric environment, certain aspects of stomach anatomy need to be highlighted. Anatomically, the stomach is divided into four regions (Figure 3-2): the cardia (gastroesophageal junction), the fundus, the body, and the pylorus (pyloric antrum and pyloric canal). Additionally, there are two other regions (the lesser and greater curvatures) that are generally not considered for tumour classification.

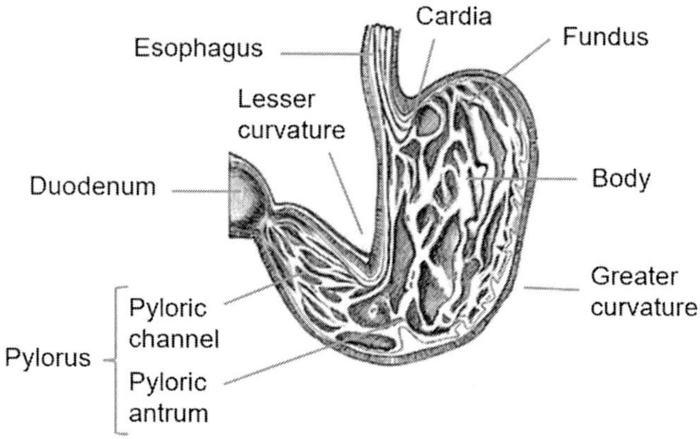

Figure 3-2. Anatomy of the stomach. Highlight of the four anatomical regions: cardia, fundus, body, and pylorus. Drawing by Daniel Ferraz (2024).

Most individuals infected with *H. pylori* develop mild pangastritis, a condition that does not adversely affect gastric physiology. Antrum-predominant gastritis presents a low risk for gastric cancer but a high risk for duodenal ulcer. In contrast, when gastritis is body-predominant, leading to hypochlorhydria (reduction of HCl) and atrophic gastritis, the risk of developing gastric cancer significantly increases.

Among the various risk factors for the development of gallbladder cancer are cholelithiasis, obesity, and exposure to certain chemicals, but chronic infection of the bile ducts by *Salmonella typhi* is also of great importance. Chronic carriers of this pathogen have an approximately eightfold higher risk of developing gallbladder carcinoma compared to non-carriers.

Reading Suggestions

The adverse consequences of not using sunscreens
International journal of cosmetic science
Antony R. Young

Human Papillomavirus Vaccines
The Journal of Infectious Diseases
Lauri E Markowitz and John T Schiller

Film Suggestion

Erin Brockovich

Steven Soderbergh's 2000 film tells the true story of Erin Brockovich, played by Julia Roberts, a persistent and determined woman who secured a favourable verdict against a polluting company for approximately 600 residents of an area affected by hexavalent chromium.

4. GENERAL OVERVIEW OF THE CELL CYCLE

Introduction

New cells can only arise from pre-existing cells; in other words, the only way to create new cells is by duplicating those that already exist. This simple fact, first established in 1855 by Rudolf Virchow, carries a profound message about the continuity of life. All living organisms result from cycles of cell growth and division, dating back to the earliest life on Earth. These cycles ensure embryonic development, organismal growth, and tissue regeneration.

The specifics of the cell cycle can vary significantly between organisms and at different stages of an organism's life. However, some characteristics are universal, and the essential sets of processes a cell must perform are those that enable it to carry out its fundamental task of transmitting genetic information to the next generation of cells. Cell division is an impressive process because the daughter cells are exactly identical in their chromosomal and genetic aspects.

Each human cell contains 46 chromosomes, each composed of a DNA molecule. During the cell cycle, these molecules are duplicated and evenly distributed to the daughter cells. This equal distribution is facilitated by the condensation of the chromosomes, allowing each one to take on a compact rod-like shape, which prevents the chromosomal strands from becoming entangled during their segregation into the daughter nuclei. At the end of division, the chromosomes decondense and return to their appearance as long DNA strands.

The mitotic capacity of a tissue decreases as its cells become more differentiated (specialized). In this way, adult multicellular organisms have highly specialized cells, such as most striated muscle cells and neurons, which no longer divide. Other cells, however, continue dividing, such as those at the base of the epidermis and the hematopoietic cells in the red bone marrow.

According to the classification proposed by Giulio Bizzozero (1846-1901), cells like hematopoietic cells are considered labile, while neurons are classified as permanent cells. There are also stable cells, which, despite being highly specialized, can regain the ability to divide if needed. For example, fibroblasts in the skin can resume division when a wound occurs, aiding in the healing process.

Cell cycle duration

The basic function of the cell cycle is to accurately duplicate the vast amount of DNA in the chromosomes and then precisely segregate the copies into the nuclei of two genetically identical daughter cells. Based on these cycle processes, cytologists define two distinct phases: cell division and interphase. Cell division includes mitosis or the M phase (division of the nucleus) and cytokinesis (division of the cytoplasm).

In general, interphase, defined as the period between two consecutive cell divisions, accounts for about 95% of the total duration of the cell cycle. In cultured fibroblasts, the cell cycle lasts around 20 hours, with interphase accounting for about 19 of those hours. The remaining 5% is dedicated to cell division, a crucial process for the growth and differentiation of the organism, transforming the zygote into the trillions of cells of the adult individual and also playing a role in regeneration processes.

It is well established that most cells require much more time to grow and duplicate their mass of proteins and organelles than to replicate their chromosomes and divide. To allow more time for growth, most cycles include additional gap phases – G_1 phase between cell division and DNA replication, and G_2 phase between replication and the next cell division.

The initial event of the cell cycle is cell growth, marked by an increase in cell mass, known as the G1 phase (G = gap). According to D. Peter Snustad and Michael J. Simmons (2013), in mammalian cells that take 24 hours to complete the cell cycle, the G1 phase can last up to 10 hours. After G1, the S phase (S = synthesis) follows, lasting about nine hours, during which DNA replication occurs.

According to the same authors, after the completion of DNA replication in the S phase, the cell enters another growth phase called the G_2 phase. During this post-replication phase, which typically lasts about four hours, preparations for mitotic division take place. Finally, the M phase, or the cell division phase, takes about an hour or slightly longer to complete.

The cell cycle step by step

At this point, it is important to clarify two cases of nomenclature variation that often arise in this context. The first involves the use of the terms "cell division" and "cell multiplication" to refer to the M phase of the cycle. Despite the apparent inconsistency, there is a plausible explanation: as cells undergo mitotic divisions, their number increases, meaning the cells multiply.

In the second case, we actually encounter a difference in nomenclature among authors. For instance, according to Bruce Alberts and colleagues (2010), cell division includes mitosis (nuclear division) and cytokinesis (cytoplasmic division), as already mentioned in this chapter. However, for others, such as D. Peter Snustad and Michael J. Simmon (2013), cell division corresponds to mitosis, encompassing both karyokinesis (nuclear division) and cytokinesis.

In all cases, the cell cycle in eukaryotes is traditionally divided into four sequential phases: G_1, S, G_2, and M. The G_1, S, and G_2 phases together make up the interphase. In a typical human cell, interphase can occupy nearly 23 hours of a 24-hour cell cycle, with just over an hour dedicated to the M phase. Cell growth occurs throughout the cell cycle, except during mitosis.

In the different phases of the cell cycle, chromosomes undergo ordered stages of condensation and decondensation. In the G_1 phase, chromatin consists of long, intertwined filaments, which are not distinguishable under a microscope as individual chromosomes. Even without direct visualization, there is evidence that chromosomes in this phase have a single chromatid.

In G_1, cells increase in size and synthesize RNA and proteins before entering the DNA replication phase. In some cases, G_1 cells enter a temporary or permanent state of quiescence, without progressing further in the cycle. Quiescent cells are in a physiological state known as G_0, which can result from a lack of mitogenic factors or a high degree of cellular differentiation.

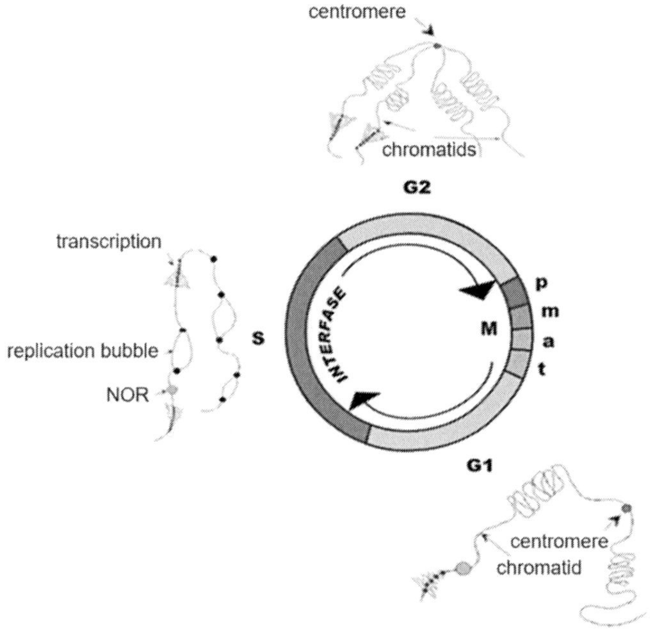

Figure 4-1. Diagram of the cell cycle. Highlighting DNA duplication in the S phase.

The S phase of the cell cycle is characterized by the replication of DNA molecules (each chromosome consists of a single, enormously long linear DNA molecule), allowing the cell to produce a complete copy of all its genes. By the end of this phase, each chromosome is composed of two chromatids joined at the centromere. The key event in chromosome duplication is DNA replication (Figure 4-1).

> A cell must address two challenges when initiating and completing DNA replication. First, replication must occur with extreme accuracy to minimize the risk of mutations in the next generation of cells. Second, every nucleotide in the genome must be copied once, and only once, to avoid the harmful effects of gene amplification (ALBERTS et al., 2010, p. 1067).

During the G_2 phase, which is shorter than G_1, there is a resumption of general protein synthesis, as well as the production of specific molecules necessary for cell division, such as the components of microtubules that will form the mitotic spindle. The onset of chromatin condensation through spiralization and the inactivation of transcription (RNA production) mark the end of the G_2 phase and the beginning of mitosis.

Mitosis is typically divided into four sequential phases. In the first phase, called prophase, chromosomes become visible due to the condensation and individualization of chromatin filaments. The nuclear envelope and nucleolus begin to fragment. The centrosomes, which have duplicated since the interphase, separate and move to opposite poles of the cell. Microtubules radiating from the poles form the aster and the spindle fibres, which help move the chromosomes.

In metaphase, chromosomes reach their highest level of condensation and take on the appearance of rod-like structures organized along the cell's equatorial plane (metaphase plate). Each chromosome, with its two chromatids, is attached to the spindle fibres via its centromere. This phase is ideal for karyotype analysis, where the chromosomes are studied in terms of their number, shape, and size.

The microtubules that make up the spindle fibres are actually attached to the kinetochore of each sister chromatid. The kinetochore is a large, multi-layered protein structure built on the heterochromatin that forms in the centromeric region of the chromosome. For accurate chromosome segregation, each kinetochore must be connected to microtubules originating from opposite poles.

During anaphase, the centromeres split longitudinally, allowing the sister chromatids to separate and migrate to opposite poles of the cell (poleward movement) as the spindle fibres shorten. Each new nucleus will receive one of the sister chromatids, now referred to as daughter chromosomes, forming a set of chromosomes identical to that existing in the nucleus of the mother cell.

From anaphase onward, the chromosomes begin to elongate slightly as they start to uncoil. Telophase begins once the chromosomes have reach the poles of the cell. The microtubules of the mitotic spindle and the aster disassemble, a nuclear membrane reforms around each daughter nucleus, and the nucleolus starts to reappear alongside each new set of chromosomes.

During telophase, the chromosomes continue to decondense and become more extended, allowing transcription to occur. The reappearance of the nucleolus in each daughter nucleus is a direct result of the transcription of ribosomal RNA molecules. Mitosis ensures that each daughter cell retains the same genetic information as the parent cell.

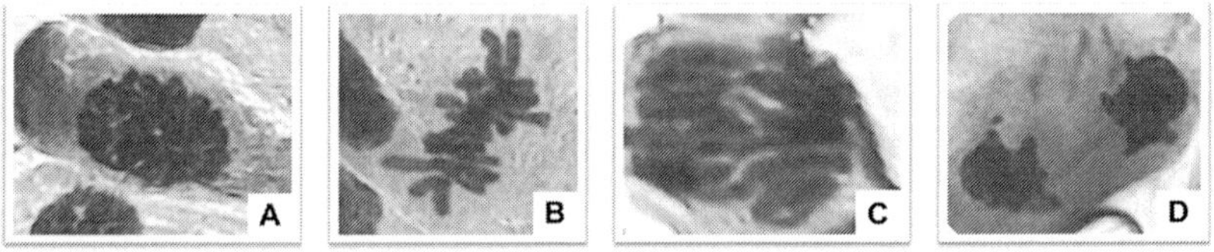

Figure 4-2. Sequential phases of mitosis. The cycle of chromosome modifications can be seen: (A) prophase, (B) metaphase, (C) anaphase, (D) telophase.

After the division of the nucleus (karyokinesis) at the end of telophase, the cytoplasmic part of the cell divides through cytodieresis (cytokinesis). In animal cells, this cytokinesis is centripetal, occurring via a median constriction that eventually separates the two daughter cells. In fact, cellular cleavage begins in anaphase with the formation of a contractile ring of protein filaments.

Cytoskeleton dynamics

The cytoskeleton is a system of protein filaments found in the cytoplasm of eukaryotic cells, providing cell shape and the ability for directed movement. The most abundant components are differentiated into three types based on the diameter of their filaments. In nanometres (1 nm = 10^{-6} mm), these are: actin microfilaments (5 to 9 nm), intermediate filaments (8 to 12 nm), and microtubules (around 25 nm).

The regulation of the dynamic behaviour and assembly of cytoskeletal filaments enables eukaryotic cells to construct a vast array of structures from their basic filament systems. A key example of the rapid reorganization of the cytoskeleton occurs during cell division, when microtubules form the mitotic spindle fibres, which are responsible for aligning chromosomes and subsequently separating them towards the poles.

Actin filaments also play a more specific role in the process of cell division. The bundles formed by these filaments contribute to the contractile ring, which pulls the membrane inward. The contraction of this structure creates a constriction in the cell, forming a cleavage furrow that gradually deepens, leading to centripetal cytokinesis. Finally, the entire filament apparatus used in cell division is dismantled, and the cell resumes its specific functions.

Reading Suggestion

Microtubules as a target for anticancer drugs
Nature Review Cancer
Mary Ann Jordan and Leslie Wilson

5. Regulation of the Cell Cycle

Introduction

The size of a multicellular organism, or of each particular organ, is primarily determined by its cell mass, which depends on both the size and the total number of cells. Naturally, this number is related to both cell production and cell death. Thus, the size of the body and each of its organs is governed by three cellular processes: mitosis, cell growth, and death.

In the microscopic world of unicellular organisms, there is a tendency to grow and proliferate at a relatively rapid pace, primarily depending on the availability of nutrients in their environment. In turn, this availability results from the total amount of nutrients and population density. However, in a multicellular organism, cell growth and division must be aligned with the needs of the organism as a whole.

The cell cycle is a universal process through which cells proliferate, ensuring the growth and development of all multicellular organisms. The most crucial events in the cycle are those that allow the copying and division of hereditary material, namely the replication of chromosomal DNA during the S phase and the separation of the replicated chromosomes during mitosis.

In general, for cells in multicellular organisms to divide, they need to be stimulated by substances called cell growth factors. Under normal conditions, cell division is triggered through signalling pathways. These pathways respond to extracellular factors, which act through a sequence of proteins in a cascade of events. For example: receptors → G protein → kinases → transcription factors.

The chemical signals that regulate the cell cycle come from both outside and inside the cell. External signals include growth factors and certain hormones. Internal signals consist mainly of two types of proteins: cyclins and cyclin-dependent kinases (CDKs). In this process, cyclins function as regulatory units, binding to CDKs, which act as catalytic units.

External signs

Cell signalling is part of a complex communication system that underpins the life of all multicellular organisms. For example, consider the signalling action of growth hormone, commonly abbreviated as GH, also known as somatotropic hormone or somatotropin, which is a polypeptide produced in the adenohypophysis (the anterior lobe of the pituitary gland).

Growth hormone binds to receptors in target tissues to stimulate growth. These receptors (GHR: growth hormone receptors) have an extracellular domain, a transmembrane portion, and a cytoplasmic domain. Upon binding with GH, the GHR undergoes conformational changes, leading to both its transphosphorylation and the phosphorylation of proteins involved in intracellular signalling.

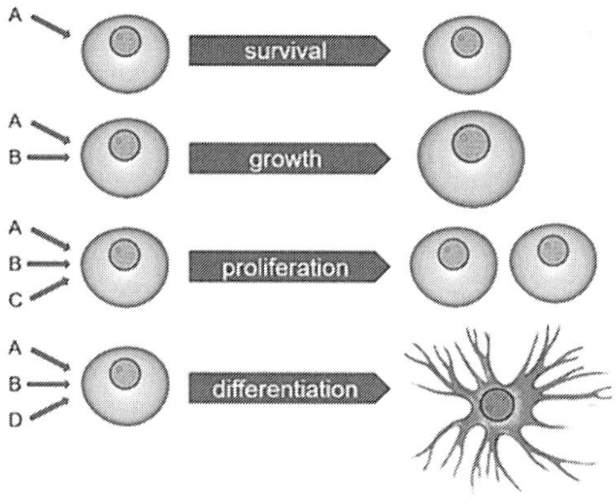

Figure 5-1. External signalling. Different combinations of signals induce different cellular responses.

Extracellular growth factors are signalling proteins that can stimulate cell survival, growth, and proliferation (Figure 5-1). The following two examples are worth highlighting: a) epidermal growth factor (EGF), which plays a key role in cell growth and proliferation; and b) platelet-derived growth factor (PDGF), which is crucial for embryonic development and tissue regeneration.

Epidermal growth factor (EGF) is a polypeptide with a molecular mass of around 6 kilodaltons (kDa), discovered in 1962 by Stanley Cohen in the submaxillary glands of mice. In humans, EGF was originally isolated from urine. It exerts a wide range of biological effects, including promoting the proliferation and differentiation of epithelial and mesenchymal cells.

Platelet-derived growth factor (PDGF) corresponds to a family of glycoproteins first identified as mitogenic agents in the blood serum of mice and monkeys. These proteins typically occur as homo- and heterodimers of two polypeptide chains, A and B, linked by disulphide bonds between cysteine residues. Since its discovery in the mid-1970s, PDGF has served as a model system for understanding how growth factors regulate biological processes.

PDGF plays a role in various cellular and molecular processes essential for tissue regeneration. Among these events, cell proliferation and extracellular matrix synthesis stand out. Due to these properties, it has long been understood that increased expression of these growth factors or their receptors is linked to the development of several neoplasms.

Signal transduction

The various molecules that function as extracellular signals typically bind to specific surface receptors on target cells, without entering the cytoplasm or cell nucleus. These receptors are proteins that act as signal transducers and fall into three main classes: ion channel-linked receptors, G-protein-linked receptors, and enzyme-linked receptors.

G proteins are a class of proteins with a crucial role in cellular signal transduction. They are named "G" proteins because they act as molecular switches, alternating between an inactive state bound to guanosine diphosphate (GDP) and an active state bound to guanosine triphosphate (GTP). Figure 5-2 below illustrates the activation cycle of a G protein.

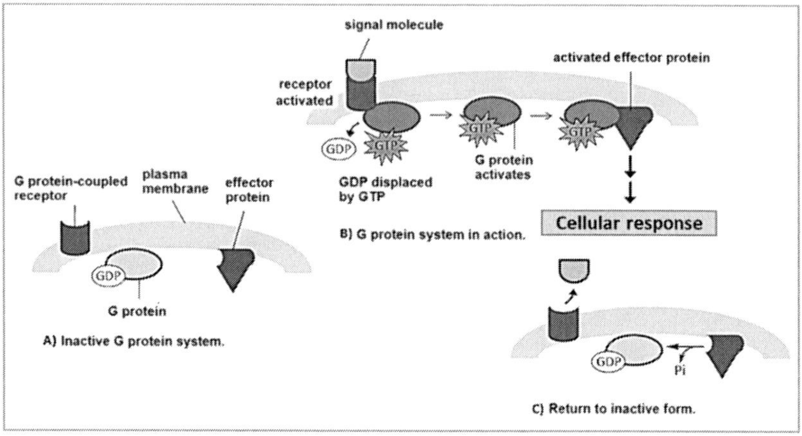

Figure 5-2. Signal transduction. Schematic of a G protein activation cycle.

To highlight the significance of knowledge in this area, researchers Alfred G. Gilman and Martin Rodbell were awarded the 1994 Nobel Prize in Physiology or Medicine for discovering G proteins in the human body. Additionally, the 2012 Nobel Prize in Chemistry was awarded to Robert J. Lefkowitz and Brian K. Kobilka for their studies on G protein-coupled receptors.

It is important to note that, even in the presence of growth factors, certain cells only replicate their DNA once they have reached a minimum size necessary to produce viable daughter cells. In fact, it was only during the 1980s that it became clear there was a control system in place, thanks to the identification of key proteins involved in regulating the cell cycle.

Internal signs

The progression of a cell through the cell cycle requires an orderly sequence of events. First, a growth factor binds to a specific receptor on the plasma membrane. Once activated, this receptor (a transmembrane protein) can activate signal transducer proteins within the cell's cytoplasm through its internal domain. Next, the signal is transmitted in a cascade through cytoplasmic transducer proteins to the nucleus, allowing the activation of nuclear regulatory proteins, which initiates and drives cell cycle progression. This intracellular signalling cascade involves the activation of effectors such as adenylate cyclases, small GTPases, phospholipases, and kinases.

CDKs: cyclin-dependent kinases. Each CDK is a protein kinase that must form a complex with a cyclin to function. Different cyclin-CDK complexes activate various stages of the cell cycle by specifically phosphorylating target proteins. In fact, the largest class of molecules involved in intracellular signalling consists of proteins that are activated or inactivated by phosphorylation.

> [...] they are, on the one hand, phosphorylated by a protein kinase, which adds one or more phosphate groups through a covalent bond, and, on the other hand, dephosphorylated by a protein phosphatase, which removes the phosphate groups from the molecule. [...] Approximately 30% of human proteins contain covalently bound phosphate, and the human genome encodes around 520 protein kinases and about 150 protein phosphatases (ALBERTS et al., 2010, p. 895-896).

Cyclins. These are proteins whose concentrations rise and fall periodically in accordance with the cell cycle. Cyclins activate essential CDKs and, therefore, help regulate the progression from one stage of the cycle to the next. Each cyclin tends to function at different stages, as shown in Figure 5-3, ensuring that the transitions between the various phases of the cycle are well-defined.

These two groups of proteins (CDKs and cyclins) are considered positive regulators of the cell cycle, as they promote the progression of the cell through the cycle, ensuring normal division into two daughter cells. However, there are also negative regulators, which inactivate the functions of the positive regulators, leading to a halt in the cycle and triggering apoptosis.

Negative regulators of the cell cycle include proteins known as cyclin-dependent kinase inhibitors (CKIs), the ubiquitin complex, and phosphatase enzymes. Proteins such as p18, p21, and p53 act as CKIs by interacting with CDKs or cyclin-CDK complexes, blocking their kinase activity. Interestingly, CDK inhibitor drugs are already being used as anticancer treatments.

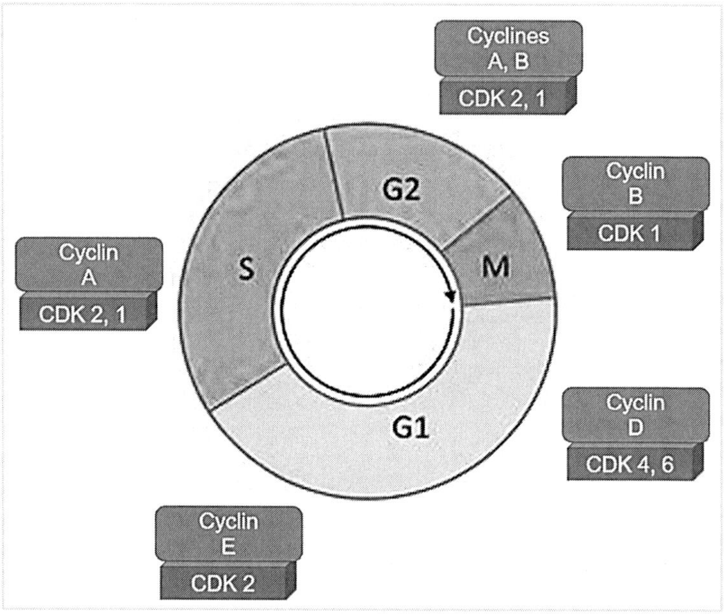

Figure 5-3. Cyclin-CDK complexes. Diagram of the cell cycle with the different complexes.

The ubiquitin complex is involved in the ubiquitination of cyclins and other proteins, preventing cell cycle progression. Ubiquitination refers to the addition of one or more ubiquitin molecules to target proteins, marking them for degradation in a proteasome. Phosphatases act by dephosphorylating CDKs and cyclin-CDK complexes, rendering them inactive.

The term apoptosis refers to programmed cell death. It comes from the Greek word "apoptwsiz", which originally described the falling of leaves from trees in autumn. The ultrastructural changes cells undergo during apoptosis were first described by Australian pathologist John Foxton Ross Kerr. Kerr and colleagues landmark paper in 1972 coined the term apoptosis to highlight its role as complementary, yet opposite, to mitosis.

Apoptosis is mediated by a family of proteolytic enzymes known as caspases. During this process, cells shrink, the nucleus fragments, and the cell surface forms blebs and breaks apart into membrane-bound fragments called apoptotic bodies. In apoptotic cells, the phospholipid phosphatidylserine moves to the outer layer of the membrane, acting as a signal for macrophages to phagocytize these cells and their fragments, all without triggering an inflammatory response.

Checkpoints

The transition from one phase of the cell cycle to another is controlled by checkpoints, mechanisms that halt the cycle until a critical process, such as DNA synthesis or the repair of damaged DNA, is completed. Figuratively speaking, would be to say that at these specific points in the cycle, the cell decides whether to proceed or pause the process for a while.

The main checkpoint in the cell cycle occurs just before the end of the G_1 phase. If conditions are favourable at this point, the cell progresses to the S phase. However, if the cell has not reached the required size (for example, due to reduced availability of nutrients in the environment) or if the necessary growth factors are absent, the cell will enter the G_0 phase.

Our fibroblasts, mentioned in the previous chapter as examples of stable cells, are skin cells, specifically from the dermis, that can remain in the G_0 phase until stimulated by platelet-derived growth factor to re-enter the cycle and repair damage from an injury. This stimulus prompts the fibroblasts to duplicate their DNA and proliferate, generating new cells to heal the wound.

The G_1 checkpoint can also detect the presence of DNA damage resulting from exposure to genotoxic agents, and in such cases, the cell cycle progression is halted before entering the S phase. Often, these damages can be repaired, allowing the cell to resume the cell cycle. However, when the damage is irreparable, the cell is directed towards apoptosis.

The G_2 checkpoint primarily serves to verify if any errors occurred during DNA replication in the S phase. Just like mentioned about the G_1 checkpoint, the presence of such errors requires proper repair or signalling for programmed cell death. The checkpoint in the M phase, on the other hand, detects potential failures in the attachment of spindle fibres to kinetochores and in the alignment of chromosomes during metaphase.

When changes in the genetic material impair the function of any of these checkpoints or prevent repairs and the apoptosis pathway, even after a normal check, such changes become mutations, as they are passed on to future generations of cells. This type of occurrence is crucial, as the accumulation of mutations is, as a rule, the main cause of cancer.

READING SUGGESTIONS

Mammalian cell cycle cyclins
Seminars in Cell & Developmental Biology
Diego Martínez-Alonso and Marcos Malumbres

Cell cycle control in cancer
Nature Reviews Molecular Cell Biology
Helen K. Matthews, Cosetta Bertoli, and Robertus A. M. de Bruin

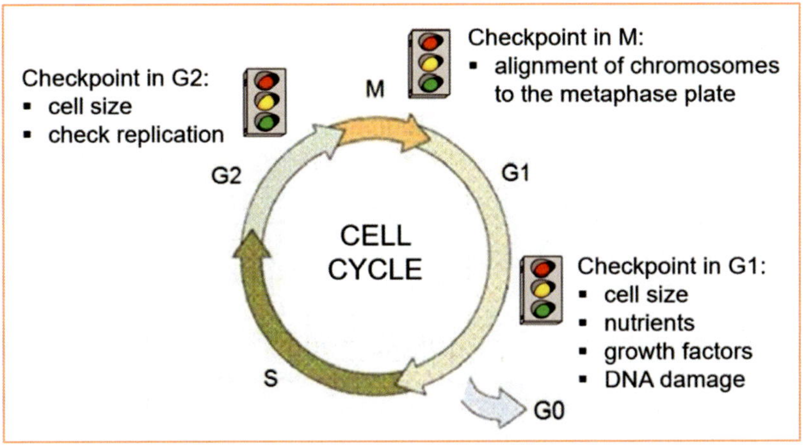

Checkpoints. Diagram of the cell cycle highlighting the most important checkpoints. Source: Rocha (2013).

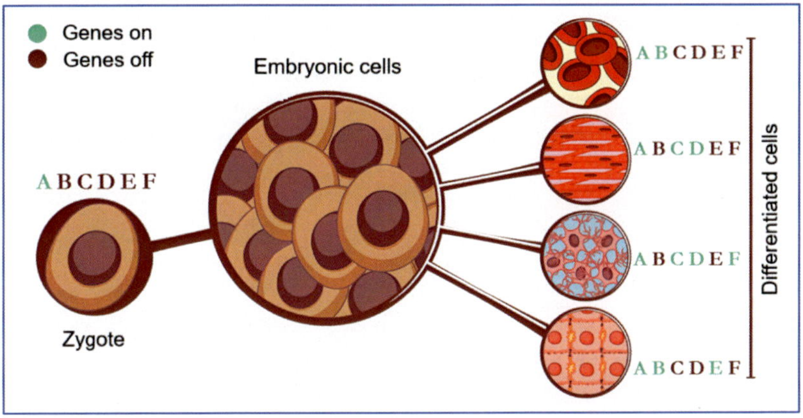

Molecular-level differentiation. The identity of each cell type is determined by the set of genes that are switched on or off. Source: Rocha (2018).

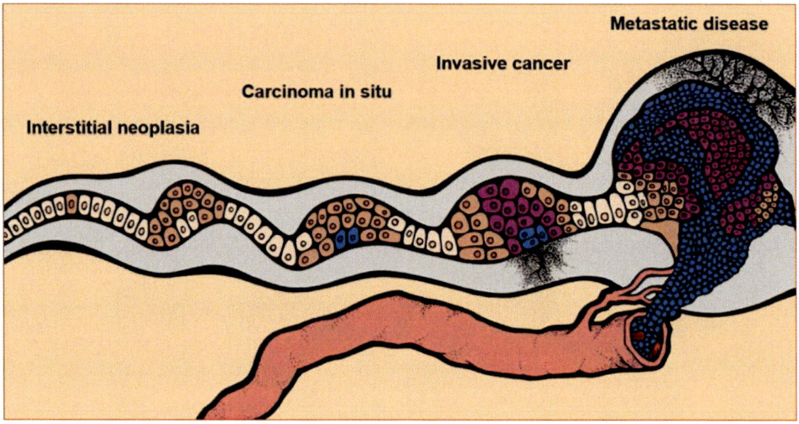

Clonal selection. Advantages for faster-growing cell variants. Drawing by Daniel Ferraz (2025).

II Meeting of the SISBIOTA-Mar Network in Vitória, Espírito Santo (Brazil), March 26th and 27th, 2012. I appear standing, checkered shirt; Sergio Floeter (Coordinator of the SISBIOTA-Mar Network); Letícia Lotufo (my supervisor in 2011-12 and Leader of the Network's bioprospecting axis).

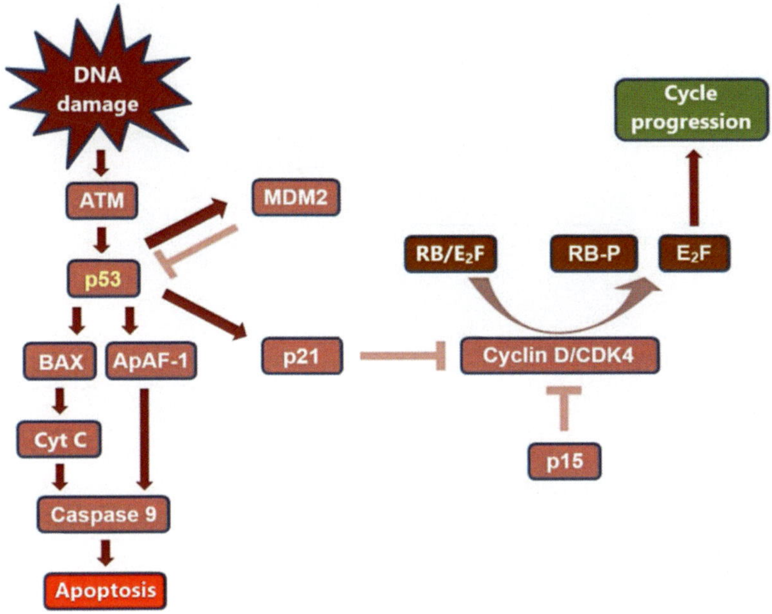

p53 Protein Interactions. A schematic representation of some functions of the *TP53* gene product and its interactions with other key proteins involved in cell cycle regulation. Own authorship (2025).

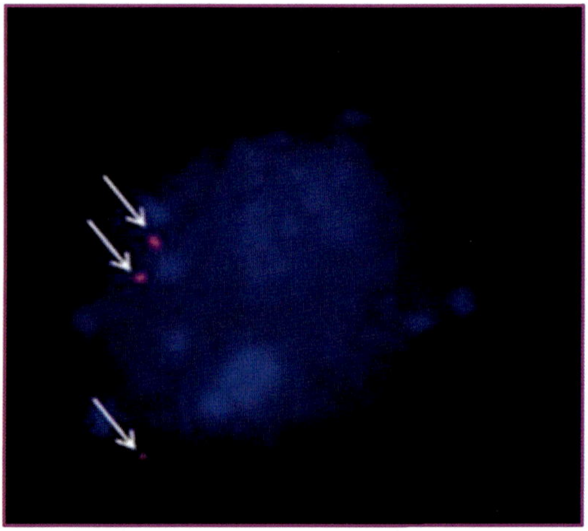

FISH analysis of a cell of PG100 line in interphase. White arrows show fluorescent signals corresponding to three copies of the *MYC* oncogene. Source: Ribeiro et al. (2010).

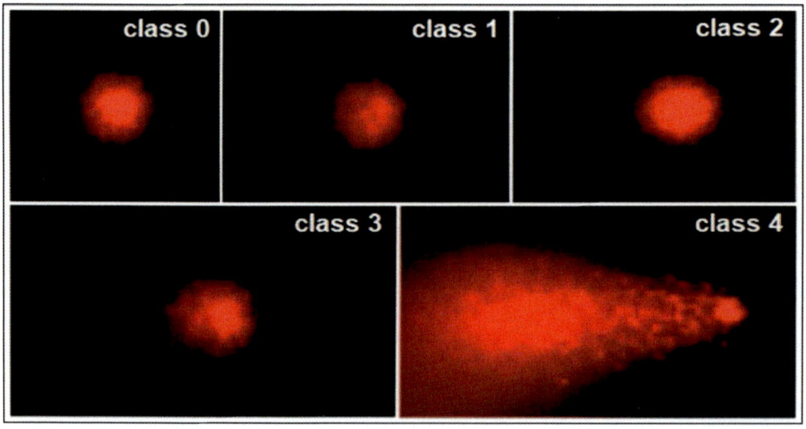

Comet Assay. Images showing the levels of DNA damage in cells. The length of the "comet tail" reflects the extent of the damage. Source: Cardoso et al. (2018).

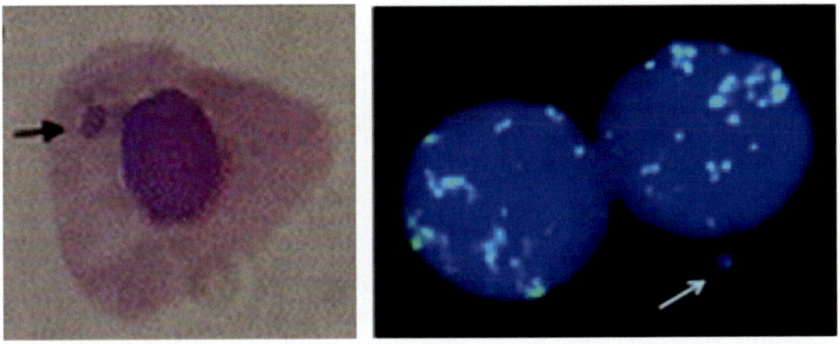

Micronucleus Test. Left: Bullfrog erythrocyte in a bioassay exposed to hexavalent chromium, with the arrow pointing to a micronucleus (Source: Monteiro et al., 2018). Right: Human blood lymphocyte after nuclear division. In this case, the micronucleus test was accompanied by FISH with a centromere-specific probe, and the arrow points to a C+ micronucleus, indicating it resulted from an aneugenic event. Source: Guimarães (2009).

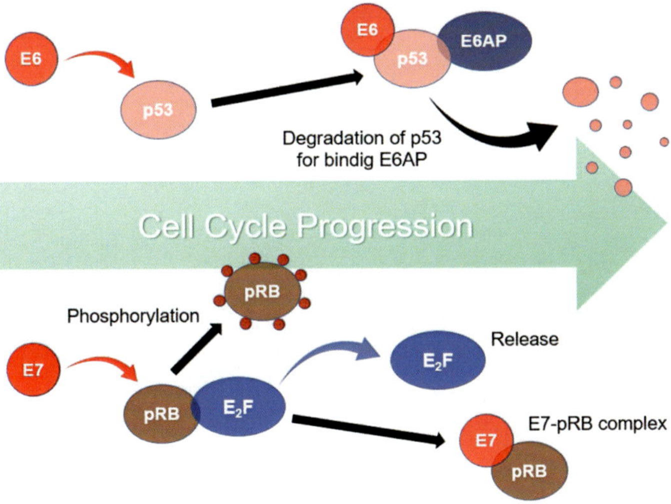

Critical Role of E6 and E7 Proteins. In the development of HPV-associated cancers, the viral proteins E6 and E7 play a crucial role in evading tumour growth suppressors. E6AP is a ubiquitin ligase. Own authorship (2025).

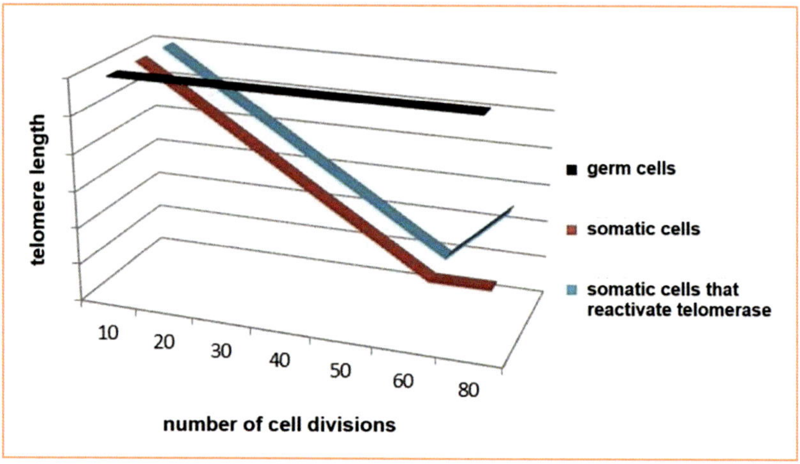

Graph showing the variation in telomere length. Cells in crisis that have already accumulated various mutations either die or reactivate telomerase, becoming immortal. Source: Rocha (2013).

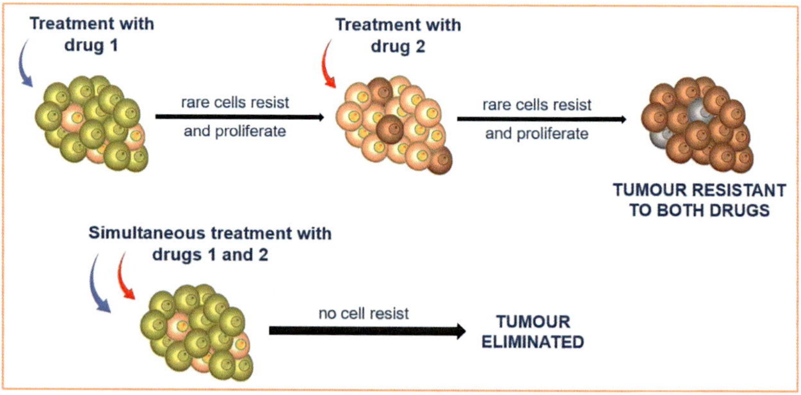

Combination chemotherapy involves the use of two or more drugs to treat a disease. In cancer therapy, treatment with multiple drugs can be more effective than sequential treatment, as it may prevent the selection of mutant cell clones resistant to both drugs. Own authorship (2025).

6. CELL DIFFERENTIATION

Introduction

Each adult individual begins as a single egg cell or zygote, with the multicellular state arising through successive mitoses, differentiation, and morphogenesis. Mitoses, along with cell volume increase, drive the growth of the body; differentiation or specialization creates cellular diversity; and morphogenesis organizes these differentiated cells into tissues and organs.

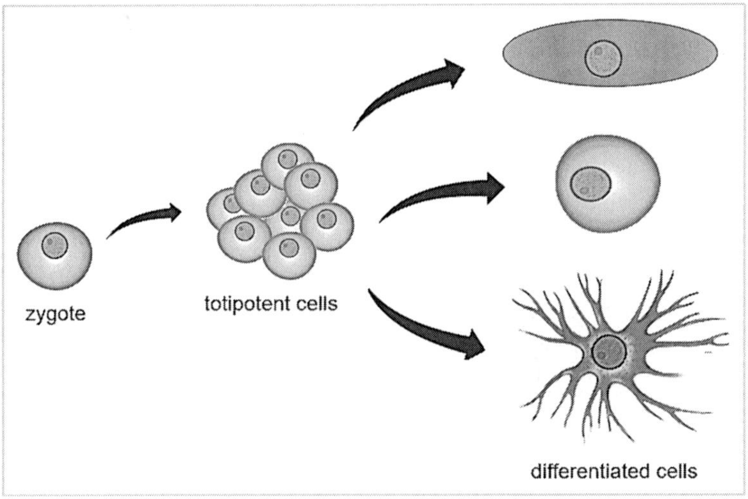

Figure 6-1. Cell differentiation. Diagram illustrating the various paths of cell specialization.

Due to the numerous and striking morphological (Figure 6-1) and functional differences among the various cell types in a multicellular organism, it becomes quite difficult to imagine that all these cells contain the same genetic material. For this reason, mainly, it was once believed that genes were selectively lost during the process of cellular differentiation.

However, advances in molecular biology indicate that as a cell differentiates, the actual changes occur in the expression of numerous genes. As mentioned in the initial chapter, the differentiation process involves the activation of specific genes, responsible for acquiring the characteristics needed by the cells of a tissue to perform their specific functions.

In fact, while some genes require activation as part of the differentiation process, others must necessarily remain active in all cells of the body, and these represent the majority of the genome. Cell types become distinct because they synthesize and accumulate different sets of gene expression products, meaning different sets of RNA molecules and proteins. Thus:

> Genes in mammalian cells can be grouped into two general classes: housekeeping genes and tissue-specific genes. Many genes encode proteins universally required to maintain the viability of all cell types throughout the body or to perform certain common biological functions [...]. These commonly expressed genes are classified as housekeeping genes (WEINBERG, 2008, p. 21).

Tissue-specific genes, although a minority in differentiated cells, are primarily responsible for the phenotypes specifically associated with each cell type. Even genes supposedly active in certain cell types do not remain active all the time. Instead, these genes are "switched on" only when their products are needed and then "switched off" again.

Among the "switched on" genes, there are some that also function to regulate other genes, and so on. Due to this interaction, one can conclude that there is a true "cascade" of gene expression regulation. This complex mechanism of expression, related to the processes of cellular differentiation, is not yet fully understood, mainly due to experimental challenges.

The expression of numerous genes, and consequently their role in cellular differentiation, is also influenced by microRNAs (miRNAs) – small RNA molecules about 17 to 22 nucleotides long that do not encode proteins. MicroRNAs play a key role in the post-transcriptional regulation of gene expression, targeting messenger RNA molecules from early embryonic development through the determination of cell lineages during organogenesis.

The importance of knowledge in this field has grown significantly in the 21st century. So much so that American biologists Victor Ambros and Gary Ruvkun were awarded the 2024 Nobel Prize in Physiology or Medicine for their discovery of microRNA and its role in gene regulation. Understanding miRNAs is crucial for gaining insight into diseases like cancer and for the development of new drugs.

On the other hand, the degree of cellular differentiation is inversely related to a cell's mitotic capacity. Thus, as cells within a tissue become more specialized in performing their specific functions, they gradually lose their ability to proliferate. Among the cells with the highest degree of differentiation and, consequently, the lowest mitotic index, are neurons and most muscle fibres.

Striated muscle fibres, for example, are so specialized that some of their components have been given distinct names. The cell membrane is called the sarcolemma, the cytoplasm (mainly filled with parallel fibrils – myofibrils) is known as sarcoplasm, the smooth endoplasmic reticulum is referred to as the sarcoplasmic reticulum, and the mitochondria are also called sarcosomes.

DNA replication and telomere shortening

An interesting aspect related to DNA replication, cell division, and the process of cellular differentiation involves the dynamics at the ends of chromosomes. These ends, called telomeres, protect against nuclease degradation, and are often compared to the plastic tips that seal the ends of shoelaces; they also prevent chromosomes from fusing with one another.

A few decades ago, it was proven that during DNA duplication, the two strands cannot be replicated in the same manner (Figure 6-2). The reason for this lies in the fact that the enzyme DNA polymerase can only synthesize DNA in the 5' ➔ 3' direction, meaning the template strands must be "read" in the 3' ➔ 5' direction. One of the strands undergoes continuous replication, however:

> [...] the second strand, or lagging strand, cannot be copied continuously because this would require DNA synthesis in the 3' to 5' direction. Instead,

> the strand [...] must be replicated in fragments: a portion of the parental helix unwinds, and a short section of the strand is replicated, then another part of the helix unwinds, and another segment of the strand is replicated (BROWN, 1999, p. 130-131).

The continuous and discontinuous strands are often referred to as the leading strand and the lagging strand, respectively. For a while, the process of discontinuous replication was only a hypothesis, which was finally confirmed in 1968 with the isolation of Okazaki fragments (Figure 6-2). These short DNA segments, ranging from 100 to 200 nucleotides in length, are named in honour of the Japanese research couple Tsuneko and Reiji Okazaki.

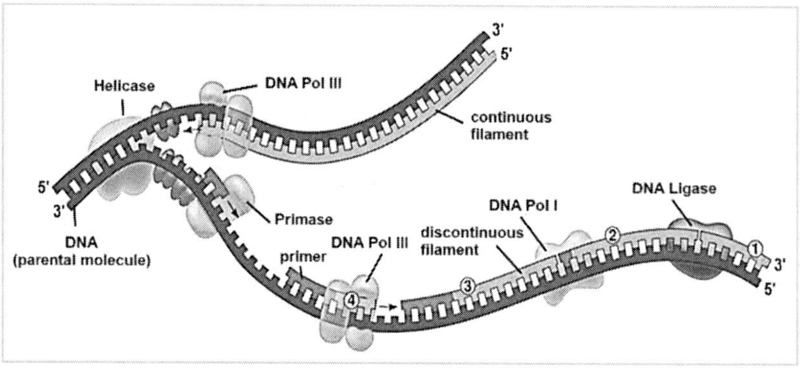

Figure 6-2. DNA Replication Fork. Diagram of continuous and discontinuous replication. Numbers 1 to 4 indicate Okazaki fragments.

To complete discontinuous replication, the adjacent Okazaki fragments must be joined, which is accomplished by phosphodiester bonds catalysed by the enzyme DNA ligase. However, at the end of the discontinuous strand (at both telomeres), there will be no DNA template strand to allow the attachment of a new primer after the last Okazaki fragment's RNA primer has been removed.

This final region, corresponding to the last primer, cannot be replicated. Thus, with each replication, the telomere shortens. Over time, the telomeres would eventually be completely lost, and losses would begin to occur in coding regions. However, the action of an enzyme called telomerase helps restore the lost telomeric fragments during replication.

Telomerase is a ribonucleoprotein composed of a protein subunit and an RNA molecule that, in vertebrates, contains about 450 nucleotides. Since this RNA component serves as a template for extending a strand of DNA, the reaction is an example of reverse transcription. Thanks to the elongation provided by telomerase, the shortening effect caused by DNA replication can be counterbalanced.

In most somatic cells, however, the telomerase gene is not "switched on," and as a result, these cells do not produce this enzyme. Consequently, with each replication cycle, chromosomes undergo progressive shortening, a phenomenon linked to replicative senescence (the limit on the number of divisions each cell type can undergo) observed in all normal human cells.

Cell differentiation and cancer

Normal cells, when cultured *in vitro*, exhibit a limited number of generations, typically ranging from 50 to 100. For example, human fibroblast cultures, the most numerous cells in connective tissue, can undergo up to 70 cycles. After this period, the culture ceases to grow, enters senescence, and eventually dies. This is not usually the case with cancer cell lines, which can continue dividing indefinitely.

Cancer cells in culture, and by extension *in vivo* neoplasms, possess the ability to divide without limits, which is why they are referred to as "immortal." The telomere of cancer cells is no longer than that of normal cells; rather, it remains stable, meaning it does not shorten during replication cycles. In fact, the telomere does shorten, but it is replenished by telomerase. The continuous activity of telomerase also occurs in most unicellular eukaryotes.

In tumour analysis, this continuous activity of telomerase is observed in 85% to 90% of cancer cells. Given that immortality is a typical feature of malignant tumour cells, the inactivation of telomerase in these cells is a possibility that is being investigated as a potential approach to control neoplastic cell proliferation. In fact, different telomerase targeting immunotherapies have been studied in pre-clinical and clinical settings.

After differentiation, the cells of each tissue acquire specific morphology and physiology, distinct from other tissues. In cases of cancer, altered cells lose their differentiation – a process known as dedifferentiation – which increases their ability to undergo mitosis, allowing them to proliferate at a faster rate than normal cells from the same tissue.

Tumour classification

> Invariably, a detailed analysis of the organization of cells within tumour masses revealed evidence of a tissue architecture that was less organized and structured compared to the surrounding normal tissues. Such histopathological comparisons provided the first seeds of an idea that would take much of the twentieth century to prove: tumours are composed of cells that have lost their ability to assemble and create tissues with normal shape and function (WEINBERG, 2008, p. 27).

In the nomenclature for benign tumours, the root of the name refers to the tissue of origin, followed by the suffix "oma." For example: osteoma, a benign tumour of bone tissue. For malignant tumours, the classification depends on the embryonic origin of the tissue first affected. For example: sarcomas, malignant tumours formed from connective (mesenchymal) tissues.

Tumours are classified based on the tissue from which they originate, although this is not always identifiable. They are categorized as epithelial tumours, connective tissue tumours, neuroectodermal tumours, hematopoietic tumours (originating from blood-forming tissues), tumours composed of multiple tissues, and tumours whose classification remains unclear.

In epithelial tissues, the nomenclature of benign tumours is based on the cells of origin, microscopic architecture, or macroscopic pattern. Adenomas are benign tumours with glandular structures, while papillomas are benign tumours that produce projections. Malignant epithelial tumours are called carcinomas, and if these neoplasms involve glandular cells, they are referred to as adenocarcinomas.

Connective tissues are highly abundant, connecting and supporting other tissues, giving structure to the body. They include a diverse family of tissues with the basic function of filling spaces and providing continuity between other tissues. There is also a wide variety of tumour types that can affect these tissues. Table 6-1 provides a summary of the main tumours affecting connective tissues.

The third major group includes tumours that develop from early nerve tissue and its associated cells. These are commonly referred to as neuroectodermal tumours due to their origin in the ectoderm of the newly formed embryo. According to Weinberg (2008), despite comprising only 1.3% of all diagnoses, they account for approximately 2.5% of cancer-related deaths.

Table 6-1. Nomenclature of benign and malignant tumours in connective tissues.

ORIGIN	TUMOURS	
	BENIGN	MALIGNANT
Fibroblasts	Fibroma	Fibrosarcoma
Mucous Tissue	Myxoma	Myxosarcoma
Adipose Tissue	Lipoma	Liposarcoma
Cartilaginous Tissue	Chondroma	Chondrosarcoma
Bone Tissue	Osteoma	Osteosarcoma
Smooth Muscle Tissue	Leiomyoma	Leiomyosarcoma
Striated Muscle Tissue	Rhabdomyoma	Rhabdomyosarcoma
Blood Vessels	Haemangioma	Hemangiosarcoma
Lymphatic Vessels	Lymphangioma	Lymphangiosarcoma

Table 6-2. Nomenclature of benign and malignant tumours in nervous tissue and related structures.

ORIGIN	TUMOURS	
	BENIGN	MALIGNANT
Glia (Gliomas)	Astrocytoma Oligodendroglioma Ependymoma	Glioblastoma multiforme
Nerve sheath	Schwannoma or Neurinoma	Malignant Neurinoma
Neurons	Ganglioneuroma	Neuroblastoma
Cerebellum	-	Medulloblastoma
Retina	-	Retinoblastoma
Adrenal medullar	Pheochromocytoma	-
Meninges	Meningioma	-

The hematopoietic tissue can be broadly classified into two types (Figure 6-3): myeloid, found in the red bone marrow, which produces red blood cells, platelets, and certain white blood cells; and lymphoid, located in organs such as the tonsils, spleen, and thymus, which produces lymphocytes, a type of white blood cell. Tumours that affect hematopoietic tissue can be classified as leukaemia or lymphomas.

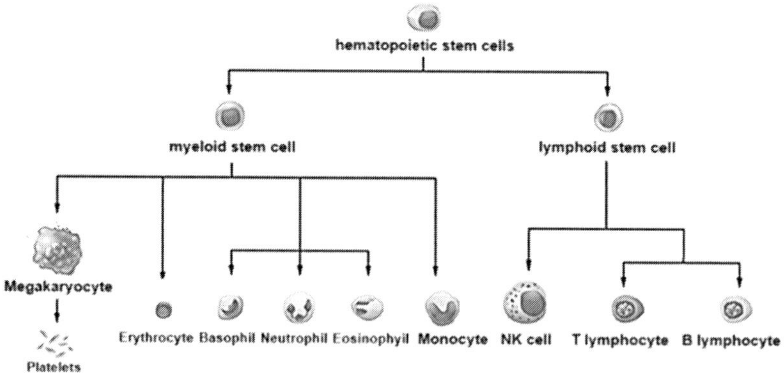

Figure 6-3. Differentiation of hematopoietic cells.

Leukaemias are malignant neoplasms of any hematopoietic variety, where the neoplastic cells lack pigmentation, and the neoplastic population is primarily found in peripheral blood. Since leukaemia does not specify the type of cell undergoing malignant transformation, it encompasses cell lineages that lead to granulocytes (eosinophils, basophils, and neutrophils) and lymphocytes.

Acute leukaemias are primary diseases characterized by an abundance of immature myeloid or lymphoid precursors (blasts): acute myeloblastic leukaemia (AML) and acute lymphoblastic leukaemia (ALL). In chronic leukaemias, the transformation occurs in more mature stem cells, and the disease tends to progress more slowly: chronic myelocytic leukaemia (CML) and chronic lymphocytic leukaemia (CLL).

Lymphomas are malignant neoplasms of lymphoid cells, more common in adults than in children, arising from the clonal proliferation of B or T lymphocytes at any stage of their maturation. In lymphomas, the transformed cells cluster together to form solid tumour masses, most often found in the lymph nodes, rather than the dispersed colonies of individual cells seen in leukaemias.

Some types of tumours do not fit into these four main groups. A notable example is what happens in the case of melanomas, which are derived from melanocytes in the skin and retina. Interestingly, these cells develop from embryonic structures called neural crests (closely related to neuroectodermal cells) but end up deviating and settle in the skin and eyes.

The World Health Organization (WHO, 2024) estimates a global burden of 35 million new cancer cases by 2050, a 77% increase compared to the 20 million estimated for 2022. This rise is attributed to population growth and aging, as well as the reduction in infant mortality and deaths from infectious diseases in developing countries, in addition to changes in population exposure to risk factors. Effectively, we are witnessing an epidemiological transition.

Table 6-3. Nomenclature of benign and malignant tumours in special tissue types.

ORIGIN	TUMOURS	
	BENIGN	MALIGNANT
Melanocytes	Melanocytic nevus (pigmented nevus)	Malignant melanoma (melanosarcoma)
Placenta	Hydatidiform mole	Chorioepithelioma (choriocarcinoma)
Ovaries	Thecoma; Luteoma	Arrenoblastoma; Dysgerminoma
Testicles	Interstitial cell tumour; Sertoli cell tumour	Seminomas
Various tissues	Benign mixed tumours; Parotid tumour; Teratoma	Malignant mixed tumours; Malignant teratoma

Primary location	cases	ASR			Primary location	cases	ASR
Trachea, bronchus, and lung	1,572,045	32.1	Men	Women	Breast	2,296,840	46.8
Prostate	1,467,854	29.4			Trachea, bronchus, and lung	908,630	16.2
Colorectum	1,069,446	21.9			Colorectum	856,979	15.2
Stomach	627,458	12.8			Cervix uteri	662,301	14.1
Liver and intrahepatic bile ducts	600,676	12.7			Thyroid	614,729	13.6
Bladder	471,293	9.3			Corpus uteri	420,368	8.4
Oesophagus	365,225	7.6			Stomach	341,326	6.0
Non-Hodgkin lymphoma	311,375	6.6			Ovary	324,603	6.7
Leukaemia	278,120	6.2			Liver and intrahepatic bile ducts	265,460	4.8
Kidney	277,800	5.9			Non-Hodgkin lymphoma	242,014	4.6

Figure 6-4. Global cancer incidence. Proportional distribution of the ten most common types of cancer estimated for 2022 by sex, excluding non-melanoma skin. Source: Prepared with data extracted from the World Cancer Research Fund (WCRF).

ASR = age-standardised rates. These are a summary measure of the rate of disease that a population would have if it had a standard age structure. Standardisation is necessary when comparing populations that differ with respect to age because age has a powerful influence on the risk of dying from cancer.

Tumours can develop in any part of the body, but some organs are more frequently affected than others. The most commonly impacted organs include the lungs, breast, cervix, prostate, large intestine (colon and rectum), skin, stomach, oesophagus, and liver. Additionally, each organ may be affected by various types of neoplasms, which can range from less to more aggressive.

Epigenetics

Epigenetics is a field of biology focused on studying changes in gene expression patterns without alterations to the primary DNA sequence. These changes result from modifications to DNA and histones, the proteins around which DNA is wrapped. Histones can undergo modifications such as methylation, phosphorylation, and acetylation. However, in the DNA molecule itself, only methylation occurs.

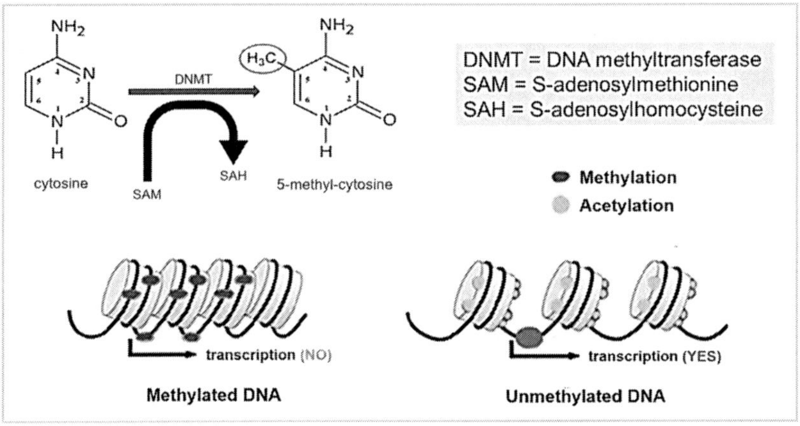

Figure 6-5. DNA Methylation. The process of DNA methylation interferes with the expression levels of various genes. Excessive DNA methylation can physically block the binding of proteins required for transcription, leading to reduced gene expression or even complete gene silencing.

According to Moshe Szyf (2007) from McGill University in Canada, methylation is a covalent modification of DNA in which a methyl group (CH_3) is transferred from S-adenosylmethionine (SAM) to the 5th carbon of a cytosine (5-MeC), which is typically followed by a guanine (CpG dinucleotide). This process is catalysed by a family of enzymes known as DNA methyltransferases (DNMTs).

The occurrence of DNA methylation can be influenced by various factors. One example is vitamin B9, also known as folic acid or folate, which plays a key role in these methyl group transfer reactions as it is essential for the synthesis of S-adenosylmethionine. Therefore, folic acid deficiency has been implicated as a risk factor for cancer, particularly colorectal cancer.

In fact, the development of nearly all types of cancer is linked to deregulated methylation events. Since actively transcribed genes tend to have low levels of methylation in their promoter regions, it can be concluded that unexpected methylation in the promoter region of a tumour suppressor gene can silence it, promoting cell proliferation and tumour progression.

On the other hand, genes that are normally silent or have low expression tend to have highly methylated promoter regions. Abnormalities that reduce methylation can also lead to the increased expression of proto-oncogenes (positive regulators that induce cell cycle progression), which may then become oncogenes (cancer-causing genes) by accelerating cell cycle events.

DNA methylation plays a crucial role in normal development and cellular differentiation in multicellular organisms, altering the expression patterns of various genes. More specifically, methylation works in conjunction with other mechanisms to ensure the appropriate level of repression of genes that are unnecessary for each cell type.

Of particular importance, for example, is the methylation's role in the dosage compensation mechanism between males (XY) and females (XX), achieved through the inactivation of a large portion of one X chromosome in female somatic cells. Finally, methylation also suppresses the expression of viral genes and other harmful elements incorporated into the genetic material of cells over the course of evolution.

READING SUGGESTIONS

T-Cell Lymphomas in South America and Europe
Brazilian Journal of Hematology and Hemotherapy
Monica Bellei, Carlos Sergio Chiattone, Stefano Luminari, and collaborators.

Establishment and conventional cytogenetic characterization of three gastric cancer cell lines
Cancer Genetics
Mariana F. Leal, José Luiz M. Nascimento, Carla Elvira A. Silva, and collaborators.

Genetics, Epigenetic Mechanism
StatPearls [Internet]
Nora M. Al Aboud, Connor Tupper, and Ishwarlal Jialal

7. MUTATIONS AND CANCER

Introduction

Cancer involves multi-stage pathologies that transform normal cells into tumour cells. In the first stage, known as "initiation," cells are exposed to carcinogens that cause changes in some of their genes. Although at this stage the cells already show some genetic alterations, and changes in their phenotype, it is still not possible to clinically detect a tumour.

The "promotion" stage is the second phase of carcinogenesis and involves changes in gene expression, selective clonal expansion, and proliferation of "initiated" cells, which are transformed into malignant cells after prolonged and continuous exposure to promoting carcinogens. If this exposure is interrupted, it is possible that carcinogenesis may also be halted at this stage.

The third and final stage is "progression," characterized by the uncontrolled and irreversible proliferation of malignant cells, which become immortal. In this stage, measurable alterations occur in the karyotype of tumour cells. Additionally, there is an increase in tumour mass, detachment of individual cells or small groups of tumour cells, and the first clinical signs of the disease emerge.

Cigarette smoke contains approximately 7,000 different chemicals, more than 70 of which are known to be carcinogenic. The carcinogenic influence of smoking can be considered complete because it contains components that act in different stages. Smoking is said to contain cancer initiators, accelerators, and promoters. Among the initiators in smoke are various polycyclic hydrocarbons, notably benzopyrene. Most carcinogens exert their effects by damaging DNA and causing mutations.

Natural selection and cancer

The first scientific explanations for the origin of life and biodiversity emerged in the early 19th century. Today, the most widely accepted version of evolutionary theory suggests that the first living beings resulted from a long and slow prebiotic chemical evolution on Earth, and that the earliest cellular organisms were unicellular and ancestral to modern bacteria and cyanobacteria.

It is believed that the protocell was surrounded by a lipid bilayer, creating a cellular compartment, and acted as a membrane, selectively permeable to certain molecules, allowing some exchanges between the interior of the protocell and the extracellular environment. In the intracellular environment there should be organic molecules dissolved in water, including energy molecules such as ATP.

> In addition, catalytic molecules (such as RNA and/or proteins) and molecules for information storage (likely RNA) would have been necessary. Clearly, the model of a cell described above is already quite complex and, therefore, could not have simply "appeared out of nowhere." The origin of this protocell was preceded by what is called chemical evolution, which is believed to have produced the basic building blocks of life [...] (MENCK & OLIVEIRA, 2001, p. 14).

Based on available information, we can assume that the first cells, dating back 3.5 billion years, were prokaryotic. However, around 2 billion years ago, eukaryotic cells emerged from some prokaryotic ones, with their interiors becoming compartmentalized, featuring various organelles and genetic material contained within the membrane-enclosed nucleus.

The formation of colonies of unicellular organisms can be observed in both bacteria and eukaryotes, such as certain protozoa and yeast. However, multicellularity is restricted to eukaryotes and most likely resulted from a long series of evolutionary events, providing advantages for body growth, diversity, cellular differentiation, and energy use.

Mutation and natural selection are the major pillars of modern evolutionary theory. Mutations represent the primary source of genetic variability in populations of living organisms, while natural selection drives the evolutionary process, favouring, over successive generations, the persistence and refinement of traits related to adaptation to particular environments.

Our decision to present this summary of the evolution of life on Earth in a work dedicated to introducing the study of cancer biology is justified by the fact that the development of cancer in an animal's body bears a strong analogy to the evolutionary process. That's right, in a way, cancer results from a selection process similar to the one proposed by Charles Darwin in 1859!

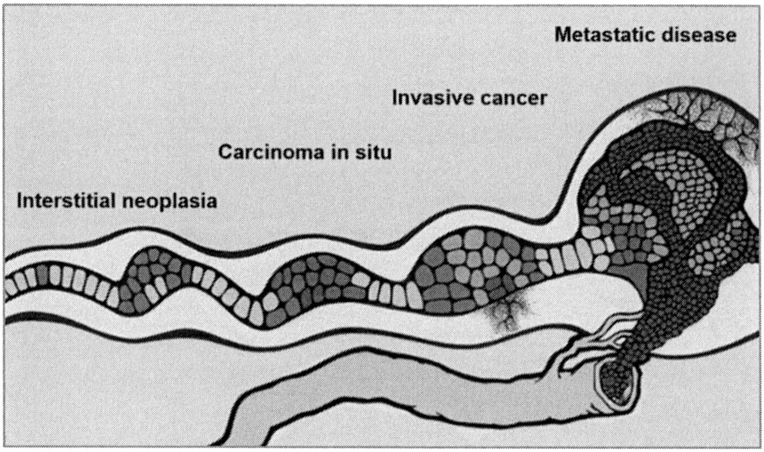

Figure 7-1. Clonal selection. Advantages for faster-growing cell variants, such that the clones within the tumour result from natural selection. Drawing by Daniel Ferraz (2024).

A tumour originates from an initial population of slightly abnormal cells that undergo multiple cycles of cell replication, accumulating mutations, and clonal expansion, so that the clones within it result from a natural selection process like in species evolution. Throughout tumorigenesis, random alterations occur, and those causing a selective disadvantage are eliminated from the tumour environment.

The cells that persist in tumour formation progressively acquire traits that distinguish them from normal cells. These traits may involve the inactivation of tumour suppressor genes or the activation of proto-oncogenes. At each stage, a cell acquires another mutation or epigenetic change that may confer a selective advantage over its neighbouring cells, making it more fit to occupy that environment.

As presented by Alberts et al. (2010), the environment within a tumour can be hostile, with low oxygen levels, nutrient scarcity, and natural barriers to growth imposed by the surrounding normal tissue. The offspring of these highly adapted cells will continue to proliferate, eventually taking over the tumour and becoming the dominant clone within the developing lesion.

Classification of mutations

Different criteria can be used to classify mutations. One of the simplest classifications is based on the type of cell in which the genetic material is altered. In this case, we refer to germline and somatic mutations. Mutations related to oncogenesis can either be inherited from the parents through the germline or arise *de novo* in the somatic lineage of a specific cell.

> Germline cells, which give rise to gametes, separate from other cell lineages early in development. All cells not belonging to the germline are somatic. Mutations in germline cells are known as germline mutations, while those in somatic cells are called somatic mutations. When a mutation occurs in a somatic cell, only its descendant cells will exhibit the mutant phenotype, and this mutation is not passed on to offspring through gametes (Adapted from SNUSTAD & SIMMONS, 2013, p. 335).

When the initial cancer-causing mutation is inherited through the germline, it will be present in all of the individual's cells, and cancer occurs as part of a hereditary cancer syndrome. In contrast, somatic mutations can occur anywhere in the body but cannot be passed on to the next generation of individuals. However, most cancers are sporadic, where mutations arise in a single somatic cell, which may then develop into a malignant tumour.

In terms of origin, mutations can be classified as either induced or spontaneous. Induced mutations result from external agents, such as radiation and certain chemicals, while spontaneous mutations arise from subtle molecular rearrangements, like tautomerization and deamination, which stem from a low level of inherent metabolic errors or the action of unknown environmental mutagens.

Mutations can also be categorized based on their adaptive value. If they introduce alleles that make the carriers more likely to benefit from natural selection, they are called advantageous mutations; otherwise, they are considered disadvantageous mutations. It is also possible for a mutation to have no impact on the organism's adaptation to its environment – this is known as a neutral mutation.

Point mutations

Changes in the DNA sequence resulting from the substitution, addition, or deletion of nitrogenous bases are known as genic or point mutations. These modifications give rise to new versions of genes – alleles – which can alter the amino acid sequence in the proteins coded by mutant alleles, potentially leading to the emergence of new traits in organisms.

Genic mutations generally result from errors during replication, affecting one or a few nucleotides in the DNA strand. In analysing genic mutations, nucleotide substitutions are classified as transitions or transversions (Figure 7-2), depending on the types of nitrogenous bases involved. It's important to note that adenine and guanine are purine bases, while thymine and cytosine are pyrimidine bases.

When a nucleotide substitution does not cause a change in the amino acid sequence of the corresponding protein, the mutation is called synonymous or silent. This is possible because the genetic code is degenerate, meaning the same amino acid can be specified by different codons. If the substitution of a nucleotide changes the amino acid sequence and functionally impairs the protein, it is known as a missense mutation.

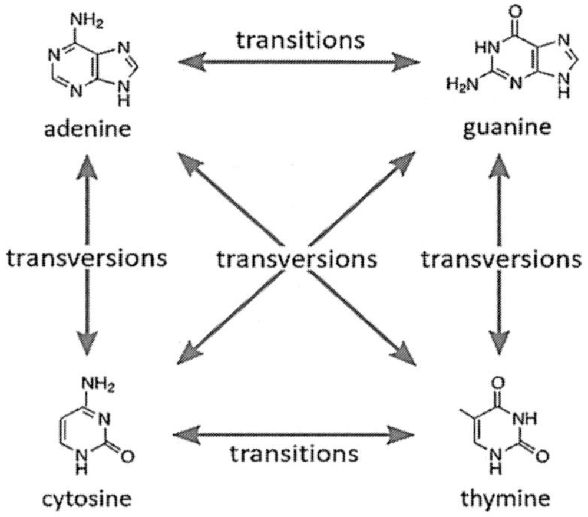

Figure 7-2. Transitions and transversions. A transition is the substitution of one purine base for another purine, or one pyrimidine base for another pyrimidine. A transversion, on the other hand, is the substitution of a purine base for a pyrimidine base, or vice versa.

On the other hand, if the substitution of a nucleotide results in the appearance of a premature stop codon in the coding strand of DNA (TTA, TAG, or TGA, corresponding to UAA, UAG, or UGA in mRNA), all amino acids coded by codons after the mutation will not be incorporated into the mutant protein, leaving it with little to no activity. This is known as a nonsense mutation.

Some genic mutations contribute to the set of alterations that accumulate during carcinogenesis. In this context, nonsense mutations in tumour suppressor genes (such as *APC*, *RB*, and *TP53*) are common. Since these suppressor genes act as negative regulators of the cell cycle, their reduction or loss of activity promotes uncontrolled cell proliferation.

> While mutations in tumor suppressor genes typically act in a recessive manner, requiring the loss of function in both alleles to drive the cell toward uncontrolled proliferation, oncogenes, by contrast, act in a dominant manner. Oncogenes result from gain-of-function mutations in just one copy of the critical proto-oncogene, which then drives the cell toward malignancy (ROCHA et al., 2024, p. 154).

The identification of cancer-related genes provides a better understanding of the disease. Today, many genes are known to contribute to carcinogenesis when altered. In addition to proto-oncogenes and tumour suppressor genes, another important group of genes involved in this process includes those related to DNA repair, which, by correcting pairing errors, prevent the establishment of mutations.

In scientific work, there are nomenclature conventions that, when known and used universally, facilitate understanding among professionals. Thus, human genes are typically written as three capitalized italic letters (e.g., *APC*, *MYC*, and *SRC* genes), while the names of their protein products are written in standard font (e.g., APC, MYC, and SRC proteins).

APC gene. A tumour suppressor gene mapped to 5p (short arm of chromosome 5), which encodes the APC protein, involved in the degradation of β-catenin (important for cell adhesion and mitosis stimulation). If the *APC* gene is mutated, it produces a truncated protein, leading to an increase in free β-catenin, which is transported to the nucleus, activating the transcription of cell proliferation genes, including *MYC*.

A mutation in the *APC* gene is often the first identifiable event in colorectal carcinogenesis. In mutant *APC*-carrying daughter cells, the loss of APC function deregulates β-catenin, stimulating cell proliferation. In addition to its association with the majority (95%) of sporadic large intestine cancers, germline *APC* mutations are responsible for triggering a hereditary cancer syndrome known as familial adenomatous polyposis (FAP).

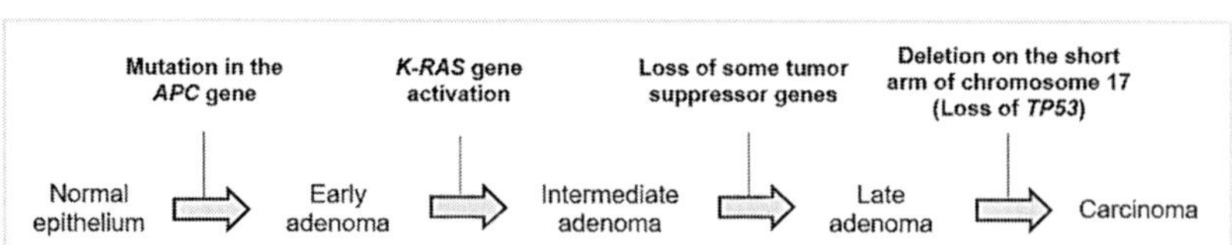

Figure 7-3. A genetic model for colorectal carcinogenesis. Mutations occurring during the progression from adenoma to colorectal carcinoma are presented.

RB Gene: A tumour suppressor gene mapped to 13q (the long arm of chromosome 13) that encodes the pRb (or RB) protein, which inhibits the cell cycle from early to mid-G_1 phase. In a normal cycle, pRb becomes inactive through excessive phosphorylation (by cyclin-CDK complexes) after the G_1 checkpoint, and from that point forward, it is fully inactivated as a suppressor, allowing cell cycle progression.

More precisely, in its non-phosphorylated or hypophosphorylated state, pRb binds to E_2F proteins (transcription factors), inactivating them. However, when pRb undergoes hyperphosphorylation in late G_1, the E_2F proteins are released to stimulate the transcription of genes under their control. These gene products drive the transition from G_1 to the S phase of the cell cycle.

There are at least two ways in which pRb function is impaired in human neoplastic cells. The first results from mutations in the *RB* gene, as seen in retinoblastomas and osteosarcomas. Another occurs in cells infected with human papillomavirus (HPV), where the viral oncoprotein E7 binds to pRb, playing a key role in the development of most cervical carcinomas.

***TP53* Gene**: A tumour suppressor gene mapped to 17p (the short arm of chromosome 17) whose transcription product is the p53 protein, a 393-amino-acid phosphoprotein. It plays a role in gene regulation, is activated by DNA damage, and assists the repair system by utilizing checkpoints to block cell cycle progression or induce apoptosis.

To date, more than 18,000 point mutations in the *TP53* gene have been described, with 97% being somatic mutations and 3% germline mutations. Approximately 50% of all tumours exhibit a loss of function in *TP53*. Unlike the *RB* and *APC* genes, where most mutations are of the nonsense type, 80% to 90% of the alterations in *TP53* are missense mutations.

Germline mutations in the *TP53* gene are responsible for a hereditary cancer condition known as Li-Fraumeni syndrome (LFS), which significantly increases the risk of cancer in affected individuals. This rare syndrome is inherited in an autosomal dominant manner, and the cancers most commonly associated with it include adrenocortical carcinomas, colorectal and breast cancers, soft tissue sarcomas, osteosarcomas, brain tumours, and leukaemias.

The significant anti-oncogenic role of the *TP53* gene has earned it the title of "guardian of the genome." The *RB* gene was the first tumour suppressor gene to be discovered, during studies on retinoblastoma. While mutations in most known tumour suppressor genes are associated only with specific subsets of human tumours, the same cannot be said for *RB* and *TP53*.

> Even after fifty tumour suppressor genes have been catalogued, pRb and p53 proteins will continue to be recognized as the products of tumour suppressor genes of prominent importance in the pathogenesis of human tumours. [...] it seems highly likely that the two signalling pathways regulated by pRb and p53 are disrupted in the vast majority of cancers (WEINBERG, 2008, p. 249).

MYC gene. A proto-oncogene mapped to 8q (the long arm of chromosome 8), involved in the regulation of the cell cycle, growth, cell adhesion, metabolism, ribosome biogenesis, protein synthesis, and mitochondrial function. It shares strong homology with cellular proto-oncogenes (c-*Myc*) from various vertebrates and also with the v-*Myc* gene from the avian myelocytomatosis virus.

The transition of the *MYC* proto-oncogene into the *MYC* oncogene plays a key role in the process of tumorigenesis in various human cancers. This activation can occur through at least four distinct mechanisms: point mutation, insertion of a provirus near *MYC*, translocations involving chromosome 8 and another chromosome (2, 14, or 22), and gene amplification, resulting in an increased number of gene copies.

Numerous proto-oncogenes can be activated into oncogenes through amplification, which is caused by DNA replication errors and can lead to the overproduction of proteins. *MYC* gene amplification has been found in gastric adenocarcinoma, colorectal carcinoma, cervical squamous cell carcinoma, and acute promyelocytic leukaemia (APL), among others.

The MYC protein represses the expression of p15, promoting the formation of the cyclin D/CDK4 complex, which is essential for the progression of the G1 phase of the cell cycle. The cyclin D/CDK4 complex phosphorylates RB, releasing E2F in the late G1 phase to activate genes that facilitate the transition to the S phase. MYC also decreases the expression of p21 and p27, freeing the cyclin E/CDK2 complex to further facilitate the G1-S transition and the formation of pre-replicative DNA complexes. Thus, MYC promotes cell proliferation by stimulating several cell cycle pathways (Figure 7-4).

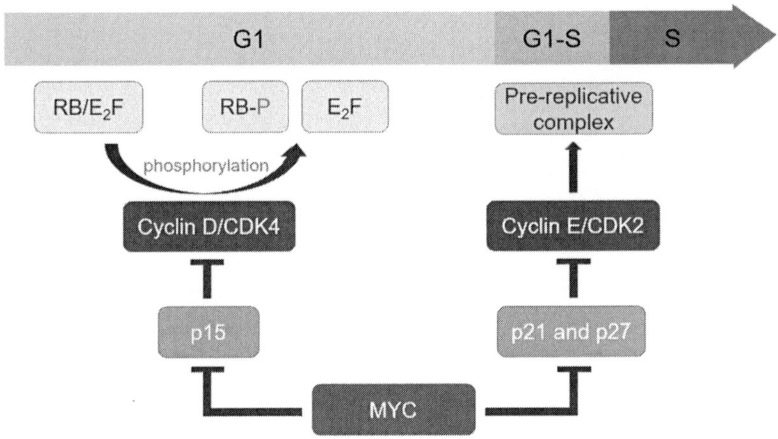

Figure 7-4. MYC gene act as positive regulators of the cell cycle. The scheme shows the pathways of action of the MYC protein by blocking CDK inhibitors.

***BRCA* Genes.** There are two genes (*BRCA1 and BRCA2*) that influence a person's likelihood of developing breast and ovarian cancer. Specifically, these genes code for proteins that help repair damaged DNA. Mutations that reduce or silence the activity of one or both *BRCA* genes significantly increase the risk of developing breast and ovarian cancer.

Chromosomal mutations

While genic mutations alter individual genes, other mutations can affect large segments of chromosomes, changing the gene sequence within a chromosome or even altering the number of chromosomes. These are known as chromosomal mutations, which can involve changes in chromosome structure or number and are usually visible under a microscope.

Structural chromosomal alterations or rearrangements generally result from breaks or fusions of chromosome segments, which are pulled by the spindle fibres (components of the cytoskeleton) during cell division. Within this group of mutations, the deletions, duplications, translocations, centric fusions, inversions, and isochromosomes are particularly noteworthy.

In a deletion, a chromosome segment is lost, leaving the "deficient" chromosome missing part of a gene or even entire genes. Meanwhile, in duplication and translocation, the chromosome gains an additional segment, with the distinction being the origin of that segment: from its homolog in duplication or from a different (non-homologous) chromosome in translocation.

Translocations can be either reciprocal or non-reciprocal, with the former being more common. In reciprocal translocations, a segment from one chromosome is exchanged with a segment from another non-homologous chromosome, resulting in two chromosomes with translocations. In non-reciprocal translocations, only one chromosome receives a segment, without transferring genetic material to the other.

In some cases, a chromosomal rearrangement visible under a microscope is repeatedly associated with a specific type of cancer. This can provide clues about the location of an oncogene that has been activated by the rearrangement (as in the case of the translocation responsible for chronic myeloid leukaemia) or a tumour suppressor gene that has been lost.

Chronic myeloid leukaemia (CML) is linked to an alteration in chromosome 22, known as the Philadelphia chromosome (named after the city in the USA where it was first discovered). This alteration results from a reciprocal translocation between chromosomes 22 and 9 (Figure 7-5). The joining of the breakpoints in this translocation creates a fusion gene (*BCR-ABL*), whose polypeptide product contributes to the development of leukaemia.

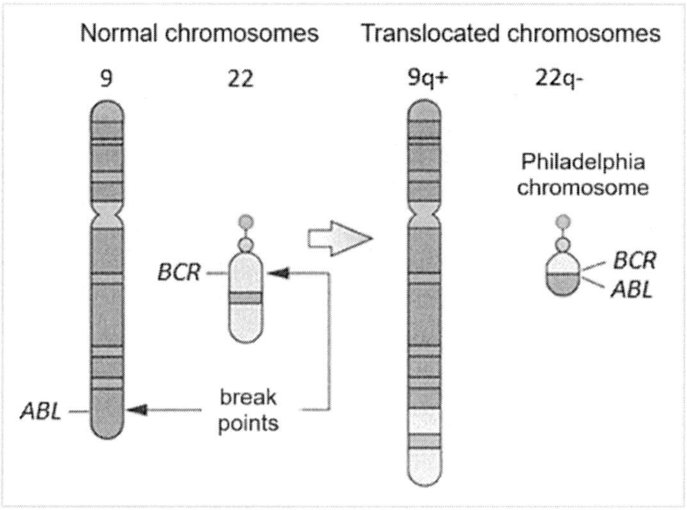

Figure 7-5. Reciprocal translocation associated with chronic myeloid leukaemia (CML). The Philadelphia chromosome (Ph) arises from a translocation involving chromosomes 9 and 22.

In an inversion, a chromosomal segment rotates 180° and reattaches. While the DNA of this chromosome retains the same nucleotides, the sequence within the inverted section is altered, affecting one or more genes. Inversions can be pericentric, occurring in a segment that includes the centromere, or paracentric, if they affect only one arm of the chromosome.

Isochromosomes result from misdivision of the centromere across the short axis (transverse division) rather than the long axis (longitudinal division), resulting in identical arms in each isochromosome. Every isochromosome is metacentric, even if the original chromosome was not. In essence, an isochromosome is a chromosome with some genes deleted and others duplicated.

Numerical chromosomal alterations or aberrations, as the name suggests, change the number of chromosomes in a cell. These are grouped into two main categories: polyploidy and aneuploidy. Polyploidy typically occurs when chromosomes are duplicated, but the cell fails to divide, affecting all pairs of homologous chromosomes. Other causes of polyploidy are cell fusion and cell cannibalism.

Polyploidization can also be a feature of carcinogenesis, as cancer aggressiveness is linked to the degree of genomic instability associated with chromosomal selection. Therefore, an increase in the population of polyploid cells may represent an intermediate stage of genomic instability during the progression of many cancer types, as the unregulated proliferation of rapidly growing cells can lead to the accumulation of multiple errors.

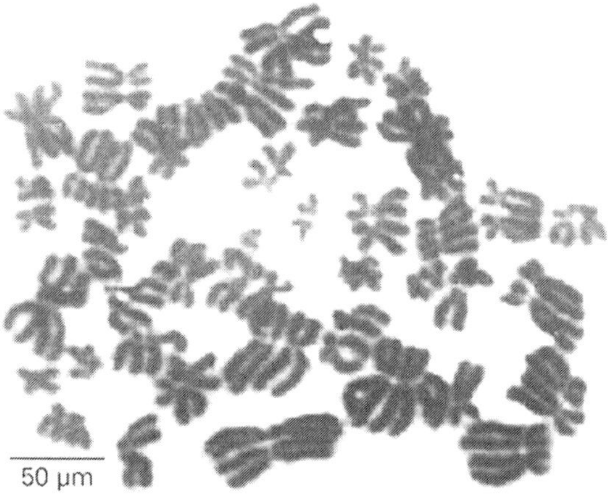

Figure 7-6. Metaphase of a polyploid human cell. Polyploidization by endoreduplication in an ACP01 cell – a neoplastic cell line established from a primary human gastric tumour. Source: Lima et al. (2004).

In aneuploidy, the cell takes on an abnormal number of chromosomes, but not an exact multiple of the haploid set of the species. This happens because only one or a few pairs of homologous chromosomes are affected, and gain or loss may occur. Aneuploidy results from errors in chromosome distribution during cell division, which can occur in both mitosis and meiosis.

Two mechanisms can cause distribution errors that result in aneuploidy. The most common is nondisjunction, where chromosomes that should migrate to opposite poles of the cell fail to separate and are pulled together by the spindle fibres. A nondisjunction event can lead to the formation of two aneuploid daughter cells: one with trisomy and the other with monosomy.

Normal human karyotypes are represented as 46,XX for females and 46,XY for males. When a karyotype is 47,XX+21 (or 47,XY+21), it means there is a trisomy of chromosome 21 (three homologs of chromosome 21, or one extra chromosome 21) – the most common cause of Down syndrome. Meanwhile, a 45,X karyotype indicates monosomy, as seen in Turner syndrome.

When we associate an aneuploid karyotype with one of these syndromes, it means that all (or most) of the patient's somatic cells exhibit this karyotype. This is not the case when it comes to the relationship between aneuploidy and neoplasms. A cancer patient typically has most of their cells with a normal karyotype, while only the tumour cells may exhibit aneuploidy and other genetic alterations.

In addition to nondisjunction, another mechanism contributing to the emergence of aneuploidy is anaphase lag. This occurs when there is a delay in the separation of sister chromatids (or homologous chromosomes) during anaphases of mitosis or meiosis. This mechanism can result in cells with an altered chromosome number and cells with a micronucleus.

The micronucleus is a small mass of genetic material enclosed by a membrane and separated from the main nucleus. It forms during telophase when the nuclear envelope is reassembled around the chromosomes. The micronucleus originates from either a chromosome fragment or an entire chromosome that lagged behind and was not incorporated into the main nucleus of a daughter cell at the end of cell division.

Molecular biology and oncological cytogenetics suggest that numerous mutations must occur for a cell to become neoplastic. Throughout the process of acquiring a neoplastic phenotype, genic and chromosomal mutations accumulate and are selected, leading to the formation of a transformed, immortalized clone capable of undergoing clonal expansion.

The karyotype of a neoplastic cell with multiple numerical (aneuploidies) and structural (deletions) chromosomal alterations is shown in Figure 7-7. This image was produced at the Human Cytogenetics Laboratory at the Federal University of Pará (Brazil), in 2010, during a cytogenetic characterization of the PG100 gastric cancer cell line.

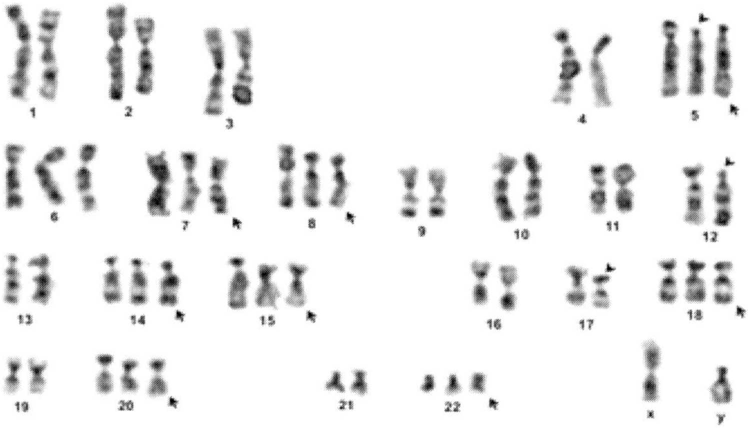

Figure 7-7. Cytogenetic analyses of the PG100 cell line. Karyogram of a metaphase. The arrows indicate trisomies of chromosomes 5, 6, 7, 8, 14, 15, 18, 20, and 22; the arrowheads indicate deletions in regions 5p, 12p, and 17p. Source: Ribeiro et al. (2010).

Environmental mutagenesis

Studies on mutations in cells, both spontaneous and induced by environmental agents, date back to the early 20th century. The results of this research emerged at a time when the scientific community was seeking to understand which groups of environmental agents could cause genetic damage. There was also growing curiosity about possible connections between mutagenic and carcinogenic agents.

> Mutagenic agents, which alter the base sequences in DNA, can accelerate [...] the appearance of mutations [...] associated with the development of neoplasms. After undergoing several divisions, a cell may accumulate mutations that, if numerous, can lead to the loss of control over its division,

> ultimately resulting in the onset of cancer (RIBEIRO et al., 2003, p. 21-22).

A wide range of scientific data supports the hypothesis that environmental factors are the primary cause of cancer, as explained by the somatic mutation theory. There is also experimental evidence that carcinogens, or their metabolic derivatives, are also mutagenic. However, this does not mean that every mutagen is carcinogenic. Copper, for example, has mutagenic potential, however it is considered non-carcinogenic.

Physical, chemical, or biological agents capable of damaging a cell's DNA are referred to as genotoxic. However, many of these damages are repaired before cell division occurs and are not permanently fixed in the genetic material. An agent is considered mutagenic only if the damage it causes to the DNA is not repaired, thus passing on to future generations of cells.

The identification and monitoring of mutagenic compounds in the environment have garnered significant attention, as these compounds can lead to an increase in cancer incidence and the development of other diseases. Research in this area includes both field studies for *in situ* biomonitoring and the use of biomarkers in controlled laboratory bioassays.

Biomarkers are an effective measure that can reflect the interaction between a biological system and an environmental agent, as evidenced by biochemical, cellular, histological, physiological, or behavioural changes. Among the numerous tests for genotoxicity and mutagenesis are the Ames test, the comet assay, the chromosomal aberration test, and the micronucleus test.

The Ames test is the most widely used method for detecting genotoxic substances. Developed by Bruce Ames and collaborators, it utilizes strains of *Salmonella typhimurium* with mutations in the locus responsible for the biosynthesis of the amino acid histidine. These bacteria (*his*-) cannot synthesize histidine and will only proliferate in a culture medium supplemented with this amino acid.

> The scientist observed that when these mutants were exposed to a mutagenic agent, the *his-* genotype could be modified and reverted to *his+*. Upon contact with the mutagenic agent, some cells would revert and begin to proliferate [...]. This reversal indicated that changes had occurred in the codons, allowing the bacterial cell to synthesize the amino acid and multiply (AIUB & FELZENSZWALB, 2011, p.11).

The comet assay is a sensitive electrophoresis technique used to analyse and quantify DNA breaks. In the results, cells without DNA damage display a round or oval nucleus, while in damaged cells, the nuclear material shows a head composed of intact DNA and a tail formed by the fragments (resembling a "comet" with a distinct head and tail). The length of DNA migration reflects the extent of damage.

The chromosomal aberration test evaluates structural alterations caused by xenobiotics in one or both chromatids of a chromosome, resulting in breaks either alone or followed by re-joining. This test can detect various aneuploidies as well as the subtle displacement of a misaligned or discontinuous fragment – larger than the width of the chromatid.

The micronucleus test can detect compounds that interfere with the formation of the mitotic spindle, disrupting the equal distribution of chromosomes to daughter cells, or affect chromosomal structure, leading to breaks. This test, therefore, detects aneugenic events (changes in chromosome number) and clastogenic events (changes in chromosome structure).

The micronucleus test was developed by Schmid in 1975, using bone marrow cells from small mammals, and has been extensively applied to assess the genotoxicity and mutagenicity of chemical compounds. Excellent results have been achieved with this test in invertebrates, fish, and amphibians, in bioassays and in the biomonitoring of contaminated areas.

READING SUGGESTIONS

Natural Selection in Cancer Biology: From Molecular Snowflakes to Trait Hallmarks
Cold Spring Harbor Perspectives in Medicine
Angelo Fortunato, Amy Boddy, Diego Mallo, and collaborators.

Mutant p53: it's not all one and the same
Cell Death & Differentiation - Nature
Margaret C. Kennedy and Scott W. Lowe.

***MYC* and gastric adenocarcinoma carcinogenesis**
World Journal of Gastroenterology
Danielle Q. Calcagno, Mariana F. Leal, Paulo P. Assumpção, and collaborators.

8 8. THE CANCER'S LEGS

Introduction

The crab is an animal belonging to the phylum Arthropoda (arthropods), subphylum Crustacea (crustaceans), class Malacostraca (malacostracans), and order Decapoda (decapods). Arthropods include crustaceans, insects, arachnids, and myriapods, which are various invertebrates characterized by jointed legs and an external skeleton primarily composed of the polysaccharide chitin.

The division of the body and the number of legs vary across different different groups of arthropods. In insects, for example, there are six legs, and the body is divided into a head, thorax, and abdomen. In arachnids, there are eight legs, and the body is divided into a cephalothorax and abdomen. In arthropods of the subphylum Crustacea, the body is also divided into a cephalothorax and abdomen, but the number of legs varies significantly across its classes and orders.

The most prominent crustaceans, such as crabs, shrimp, and lobsters, belong to the order Decapoda, a name that logically relates to their number of legs (*deca* = ten + *podos* = legs, feet, or limbs). In zoological terms, however, these ten appendages attached to the cephalothorax of decapods, which are used for movement on surfaces, are referred to as pereiopods.

As mentioned in the first chapter, the malignant cells of a tumour proliferate and infiltrate between healthy cells in a manner similar to the legs of a crab digging into the sand. This image inspired the use of the word "cancer," which means crab. Interestingly, the number of a crab's legs is exactly the same as the number of "hallmarks" currently attributed to neoplastic growth.

The 10 "hallmarks" of cancer are traits that give certain cells advantages over others, which become selected throughout the process of tumour evolution. The book’s title, "The Crab's Legs," refers to this coincidence:

the crab's 10 legs and the 10 "hallmarks" of cancer. For the same reason, we decided to condense the content of this *Introduction to Cancer Biology* into 10 chapters.

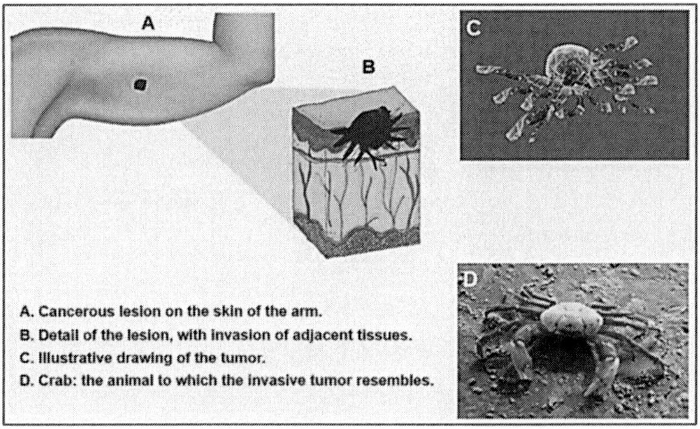

Figure 8-1. Crab and tumour. Tumour cells proliferate and invade between healthy tissue cells as if they were the legs of a crab.

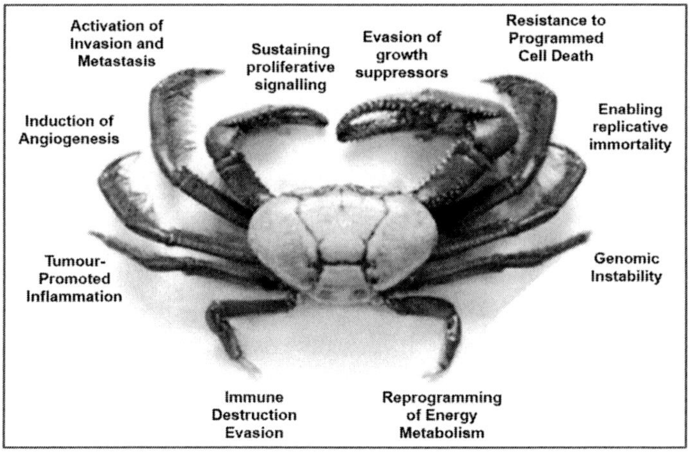

Figure 8-2. The 10 "hallmarks" of cancer. Physiological changes that lead to tumours.

The "hallmarks" of cancer

> Cancer cells are defined by two inheritable properties: (1) they reproduce without adhering to the normal limits of cell division, and (2) they invade and colonize regions typically reserved for other cells. It is the combination of these two activities that makes cancer particularly dangerous. An abnormal cell that grows (increases in mass) and proliferates (divides) uncontrollably will form a tumour or neoplasm – literally, a new growth (ALBERTS et al., 2010, p. 1206).

In fact, the properties of cancer cells must be understood in detail. Just over two decades ago, researchers had already identified six fundamental alterations in cellular physiology that collectively lead to malignant tumours: (1) sustaining proliferative signalling, (2) evading growth suppressors, (3) resisting programmed cell death, (4) enabling replicative immortality, (5) activating invasion and metastasis, and (6) inducing angiogenesis.

The six alterations were masterfully summarized in Douglas Hanahan and Robert Allan Weinberg's article "The Hallmarks of Cancer," published in January 2000, which became one of the most significant publications in the field. The vast catalogue of cancer cell genotypes is a manifestation of these six alterations, with each physiological change (an ability acquired during tumour development) representing the breakdown of an anti-cancer defence mechanism in cells and tissues. The authors suggested that most, if not all, cancers must acquire the same set of functional capabilities during their development, although through varied strategies or mechanisms.

1. Sustaining proliferative signalling. Under normal conditions, active cell proliferation can only occur following stimulating signalling events from mitogens, such as growth factors. However, during carcinogenesis, many activated oncogenes are capable of mimicking normal growth signalling. During tumour formation, proliferative signalling is sustained by the deregulation of signals that control and promote the growth of normal cells, interfering with the cell cycle and division, thereby disrupting homeostasis.

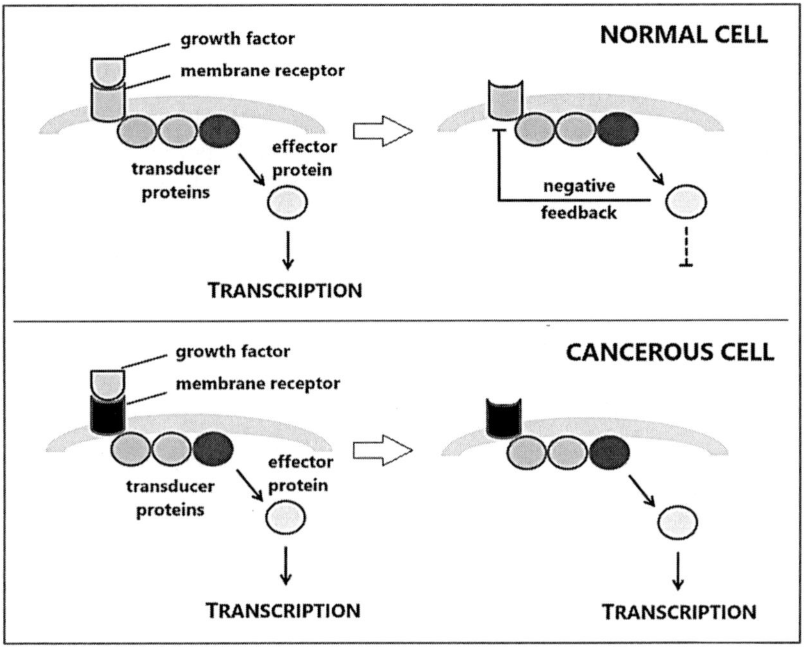

Figure 8-3. Dysregulated signalling in carcinogenesis. In the normal cell (top), signalling cascades reach the effector protein (a transcription factor) only in the presence of a growth factor. In the cancerous cell (bottom), the receptor, altered due to a mutation in the responsible gene, cannot be "switched off" and continues signalling, leading to constant activation of the effector protein and transcription.

A cell's self-sufficiency in growth factors can result, among other things, from autocrine stimulation (many cancer cells acquire the ability to synthesize their own growth factors) or from the hyperactivity of cell surface receptors that transmit growth-stimulating signals into the cell. These receptors can become targets for dysregulation during tumour pathogenesis.

Cancer cells can also alter the types of extracellular matrix receptors (integrins) they express, favouring those that promote growth signaling. Moreover, the relative intensity of oncogenic signalling may reflect a balance between maximum mitogenic stimulation in cancer cells and the evasion of antiproliferative defences. Alternatively, some neoplastic cells may adapt to high levels of oncogenic signalling by disabling their senescence and apoptosis-inducing pathways.

2. Evasion of growth suppressors. Multiple antiproliferative signals operate in normal tissues to maintain cellular quiescence and tissue homeostasis. These signals include soluble growth inhibitors and immobilized inhibitors embedded in the extracellular matrix. Emerging cancer cells must evade these antiproliferative signals to continue proliferating.

In the G_1 phase, cells monitor their external environment and, based on the signals detected, decide whether to proliferate, remain quiescent, or enter an advanced state of differentiation. Deprivation of growth factors or nutrients, signalling through extracellular matrix molecules, as well as DNA damage, can induce cell growth arrest, halt proliferation, and trigger apoptosis.

At the molecular level, many, if not all, antiproliferative signals are channelled through the pRB protein, a negative regulator of the cell cycle in G_1, which we discussed in the previous chapter. Disruptions in the pRB pathway, caused by mutations in the *RB* gene, release E_2F and allow progression from the G_1 to the S phase, even in the presence of proliferation suppressors.

An interesting situation is observed in HPV-associated cervical cancer. The cellular proteins p53 and pRB are the most important gene products in tumour growth suppression. On the other hand, the HPV proteins E6 and E7 are essential for oncogenic cellular transformation. E6 mediates the degradation of p53, and E7 can form complexes with pRB, impairing its function. Thus, we see that virus-infected cells can evade tumour suppressors due to the action of these two viral oncoproteins.

3. Resistance to Programmed Cell Death. In the absence of cancer, cells with molecular damage undergo apoptosis and are removed by specialized phagocytes at the end of the process, in an efficient, inflammation-free mechanism of programmed cell death. This mechanism is mediated by enzymes called caspases and serves as a crucial barrier to tumour development.

However, in malignant cell growth, certain pathways are altered, allowing transformed cells to evade apoptosis. Examples of these alterations include damage to death receptors that transmit apoptosis-inducing signals. The functional inactivation of the *TP53* gene is another mechanism that enables tumour cells to escape the process of programmed cell death.

Most mutations in *TP53* result in a non-functional protein. The inactivation of *TP53* through mutations typically occurs as a late event during the transition from adenomas to carcinomas. It is well known that cells with inactivated p53 often exhibit cell cycle disruption, impaired DNA repair, a tendency to accumulate gene mutations or chromosomal rearrangements, and resistance to apoptosis.

4. Enabling replicative immortality. Normal cells divide approximately 50 to 100 times before reaching replicative senescence. However, some rare mutations and epigenetic events can lead cells to become immortalized. One example is the activation of the telomerase gene, which is usually "switched off" in most somatic cells. Many immortalized cells give rise to cancer.

Most likely, the progression of incipient neoplasms toward becoming fully cancerous tumours is halted by senescence, apoptosis, or senescence followed by apoptosis. In rare cases where cellular immortalization occurs, it is generally attributed to the ability to maintain telomeric DNA at a length sufficient to sustain successive mitotic cycles. This condition is most often achieved through the upregulation of telomerase expression.

5. Activation of Invasion and Metastasis. During the development of most cancers, primary tumours produce pioneer cells that move outward, invading adjacent tissues, and then travel to distant sites where they can successfully establish new colonies. As argued by Michael B. Sporn in his influential 1996 article, "The War on Cancer," local invasion and distant metastasis from the initial tumour are far more lethal than the cell proliferation itself.

When a malignant tumour reaches about 1 cm^3, it becomes detectable by available diagnostic methods and typically contains around 10^9 cells. The neoplastic formation undergoes different rates of doubling at various points in its natural history, and in many cases, metastasis – via the bloodstream or lymphatic system – has likely already occurred well before detection.

> In Laurén's classification, intestinal-type gastric carcinoma is characterized by being well-differentiated histologically, more frequent in older males, and commonly associated with gastric mucosa displaying intestinal metaplasia. Its spread is almost always hematogenous. On the other hand, diffuse-type gastric carcinoma is marked by a loss of cell adhesion, shows no gender preference, is more frequent in younger individuals and in countries with lower stomach cancer incidence. It spreads rapidly through the submucosa and via the lymphatic pathways (ROCHA et al., 2024, p. 153).

Metastases are the final products of a multi-step process known as the invasion-metastasis cascade, which involves the spread of local cancer cells to distant organs from the primary tumour, followed by their adaptation to different microenvironments. Each of these events is driven by the acquisition of genetic and/or epigenetic changes within the tumour cells.

6. Induction of Angiogenesis. The formation of new blood vessels is essential for tumour growth and metastasis. Tumours require a blood supply for oxygen and nutrients to reach a significant size and, subsequently, to invade distant sites. As a result, these neoplastic masses secrete pro-angiogenic growth factors. During tumour progression, an "angiogenic switch" is almost always activated and remains on, causing new vessels to continuously sprout from the existing vasculature, helping to sustain the expansion of neoplastic growths.

In this process, two of the most important secretions are vascular endothelial growth factor (VEGF) and fibroblast growth factor (FGF), whose increased expression can be triggered by reduced oxygen tension (hypoxia). VEGF refers to a family of protein isoforms directly involved in angiogenesis, with the ability to promote the growth of endothelial cells derived from arteries and veins. FGF, released by fibroblasts, platelets, and macrophages, has also been shown to have mitogenic and angiogenic activity.

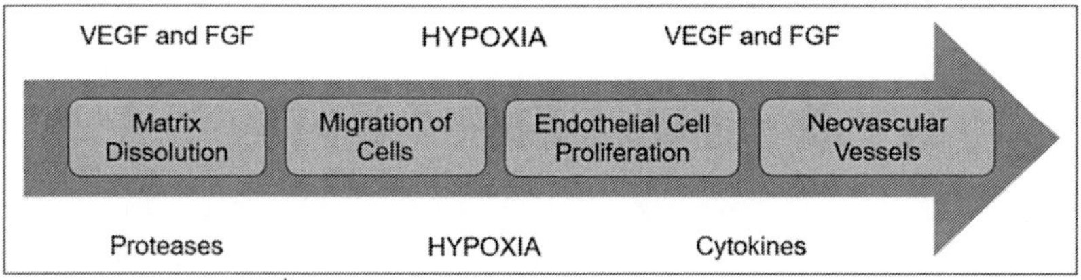

Figure 8-4. Induction of angiogenesis. When an early-stage tumour begins the process of angiogenesis, hypoxia signals induce the production of angiogenic growth factors. As the process advances, the new blood vessels enable the tumour to fully develop and spread to other organs.

The understanding of these six alterations already allowed for a solid characterization of neoplasms. However, much more remained to be discovered. With optimism, in their first work on the "hallmarks" of cancer, Hanahan and Weinberg (2000) predicted that within 20 years, the circuitry of a cancer cell would be mapped as completely as the components of a computer chip.

In 2011, Hanahan and Weinberg published a new article: "Hallmarks of Cancer: The Next Generation." In this work, they expanded the list to a total of 10 hallmarks. According to the authors, underlying the six previously discussed traits are: (7) genome instability, which accelerates the acquisition of genetic diversity, and (8) inflammation, which promotes multiple hallmark functions.

It became evident that, in addition to cancer cells, tumours create a tumour microenvironment, consisting of normal cells, blood vessels, soluble factors, and other molecules. Conceptual progress over just a decade added two emerging hallmarks to the list: (9) reprogramming of energy metabolism and (10) the ability to evade destruction by the immune system.

7. Genomic Instability. The genetic alterations that drive tumour progression largely result from genomic instability. One primary source of this instability is the loss of telomeric DNA in many tumours, which leads to karyotypic instability. This is often associated with chromosomal rearrangements, such as amplifications and deletions of segments across different chromosomes.

Various defects affecting several components of the DNA maintenance machinery – also known as the "caretaker genes" of the genome – have been documented. The catalogue of defects in these genes includes those whose products are involved in detecting DNA damage and activating repair mechanisms, directly repairing damaged DNA, and inactivating or intercepting mutagenic substances before they can harm the DNA.

When a cell sustains damage beyond its repair capacity, it is subjected to severe alterations such as chromosomal translocations, segment deletions, amplifications, duplications, or inversions, as well as base pair substitutions in the DNA. The variability observed in tumour cell clones is a result of the genomic instability that typically arises throughout the process of tumorigenesis.

8. Tumour-Promoted Inflammation. Some tumours are densely infiltrated by immune cells, much like what is seen in inflammatory conditions in non-neoplastic tissues. Moreover, recent studies have shown that the progression of disease into an invasive form of cancer often requires cells that normally participate in wound healing processes. In this way, inflammation becomes an accomplice in carcinogenesis, contributing to disease progression by supplying bioactive molecules to the tumour microenvironment, including growth factors that sustain proliferative signalling, survival-promoting factors, and enzymatic modifications to the extracellular matrix that facilitate angiogenesis, invasion, and metastasis.

Many chronic non-neoplastic inflammatory conditions can serve as predisposing factors for cancer pathogenesis. In the case of malignant tumour development, various populations of inflammatory cells may be detected within the tumour. On the other hand, understanding the biochemical details of inflammation and its specific contributions to tumour promotion can significantly aid in the development of anticancer drugs.

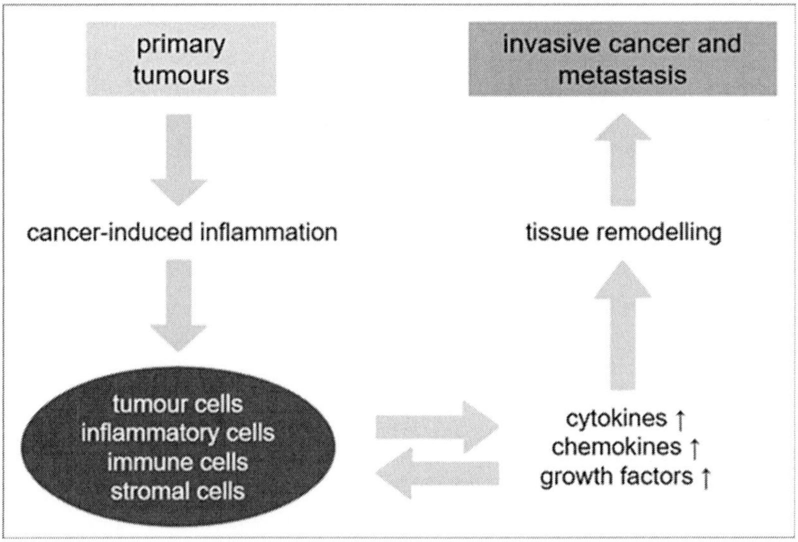

Figure 8-5. Inflammation and Cancer. Inflammation prompts tumour cells, inflammatory cells, immune cells, and stromal cells to secrete cytokines, chemokines, and growth factors. These molecules, in turn, promote the recruitment and phenotypic alterations of immune cells. Inflammation reshapes tumour tissues and creates a microenvironment conducive to the survival of cancer cells.

9. Reprogramming of Energy Metabolism. In most normal tissues, there is a metabolic program that fuels the physiological activities of the collective cells. However, during tumour development, this program is replaced or, more accurately, undergoes significant reprogramming of cellular energy metabolism to support continuous cell growth and proliferation.

Glycolysis is used in all tissues to break down glucose, providing energy in the form of ATP and intermediates for other metabolic pathways. Under normal conditions, pyruvate is the final product of glycolysis. In cells with functioning mitochondria and an adequate supply of oxygen, this aerobic glycolysis leads to the oxidative decarboxylation of pyruvate into acetyl-coenzyme A (acetyl-CoA).

According to Richard Harvey and Denise Ferrier (2012), the acetyl-coenzyme A complex is the primary fuel for the citric acid cycle (Krebs cycle) within mitochondria. However, alternatively, pyruvate can be reduced by NADH to form lactate, regenerating NAD^+ in the process. This conversion of glucose into lactate is called anaerobic glycolysis, as it can occur without the involvement of oxygen.

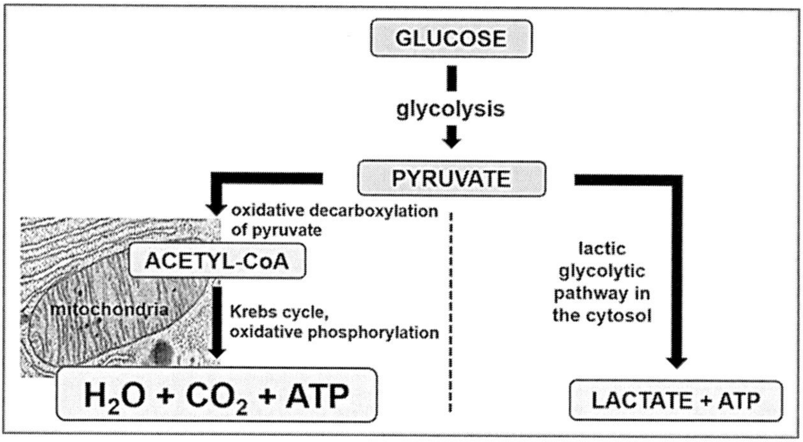

Figure 8-6. Summary of glucose metabolism. Mitochondrial and cytosolic pathways, highlighting the difference in energy yield (ATP) between them.

In the mid-20th century, German physician and physiologist Otto Heinrich Warburg was the first to observe an abnormal characteristic of energy metabolism in cancer: even in the presence of oxygen, cancer cells can reprogram their glucose metabolism and, consequently, their energy production. This process is largely limited to glycolysis in the cytosol.

The existence of this metabolic switch in cancer cells has been increasingly substantiated in recent decades. Although somewhat paradoxical, the reprogramming of energy metabolism is now well established. This means that cancer cells must compensate for the roughly 18-fold lower efficiency of ATP production through glycolysis compared to mitochondrial oxidative phosphorylation.

The glycolytic pathway is significantly less energy-efficient compared to oxidative phosphorylation. However, it allows neoplastic cells to produce intermediate molecules that serve as substrates for the synthesis of amino acids and nucleotides. As a result, the reprogramming of energy metabolism enables malignant cells to adapt and survive in low-oxygen environments while also enhancing their ability to proliferate.

10. Immune Destruction Evasion. According to the theory of immune surveillance, cells and tissues are constantly monitored by an ever-alert immune system. This immune surveillance is responsible for recognizing and eliminating various infectious agents as well as the vast majority of incipient cancer cells, and possibly even small tumours and micrometastases.

It is now known that the potential of cytotoxic T lymphocytes (CD8+), which target tumour antigens, represents one of the main effective mechanisms of antitumor immunity. This has been explored in several completed and ongoing studies. For their activation and proliferation, CD8+ cells depend on cytokines produced and secreted by helper T cells (CD4+).

Without abandoning this logic, it appears that solid tumours have developed mechanisms, not yet fully understood, that allow them to evade detection by both branches of the immune system (innate and adaptive). At the very least, they have managed to avoid eradication by limiting the typically destructive action of the immune system.

The ten "hallmarks" of cancer are mostly activated and sustained to varying degrees through the contributions of different mesenchymal cell subtypes. The roles of these cells in supporting the "hallmarks" are becoming increasingly well understood, as is their reciprocal relationship with neoplastic cells. This deeper understanding presents promising new targets for cancer therapy.

Reading Suggestions

Hallmarks of cancer: the next generation
Cell
Douglas Hanahan and Robert A. Weinberg

Telomeres, Telomerase and Cancer
Archives of Medical Research
Peter M. Lansdorp

Immuno-oncology: understanding the function and dysfunction of the immune system in cancer
Annals of Oncology
Olivera J. Finn

9. PREVENTION, DIAGNOSIS, AND TREATMENT

Introduction

The act of caregiving, as a gesture and expression of solidarity, aims to protect and ensure life, rights, and citizenship. But in order to care, one must first know. Knowledge differs from mere information because it carries intentionality. Information can even be seen as the raw material of knowledge. Both consist of true statements, but knowledge can be seen as information with a purpose or utility.

In various areas of human activity, the motivations for seeking greater knowledge can be quite diverse. However, one of two main reasons is often present. The first is to learn about things that are either useful, beautiful, or both. In this sense, knowledge can lead to the profits gained from exploiting a new and valuable resource or the enjoyment that comes from contemplation.

The second major reason is to learn about harmful things. In this case, knowledge enables the creation of both preventive and remedial measures. This type of motivation logically drives us when it comes to cancer. It is essential to understand it in order to prevent it, and, most importantly, to diagnose it early, increasing the chances that treatment will lead to a cure or control of the disease.

Careful prevention, early diagnosis, and rigorous treatment significantly improve the chances of fighting cancer. In general, good practices that promote overall health also contribute to the prevention of neoplasms. Early diagnosis is critical for a higher chance of cure, and the treatment of cancer patients must be multidisciplinary, involving collaboration between clinical specialties and, in many cases, surgical interventions as well.

Cancer prevention

In 2005, the Union for International Cancer Control (UICC) established World Cancer Day, celebrated every year on February 4th. The aim of this day is to draw the attention of nations, government leaders, healthcare managers, and the general public to the growing cancer crisis, which has reached catastrophic proportions worldwide, posing a threat to future generations.

The theme for World Cancer Day, spanning three years (2022 to 2024), is "Close the Care Gap," recognizing that each of us has the power to make a difference. Whether our individual contributions are big or small, together we can truly help reduce the global impact of cancer. The campaign is about acknowledging the strength of collective action, while understanding and addressing the inequalities in cancer care around the world.

Figure 9-1. Logo of World Cancer Day, celebrated on February 4th. From 2022 to 2024, the campaign theme suggested by the UICC was "Close the Care Gap." This day offers an opportunity to spread information and foster discussion through media outlets worldwide.

Since cancer treatment is not yet effective in all cases, prevention remains our most powerful tool for intervening in the complex process of carcinogenesis and preventing the development of tumours. To draw attention to the importance of discussing the disease and establishing prevention policies, World Day of Fight Against Cancer was instituted on April 8th.

The international movement known as Pink October is celebrated worldwide. The movement began in the United States and dates back to the last decade of the 20th century. The name refers to the pink ribbon, a global symbol of the fight against breast cancer, encouraging the participation of the public, businesses, and various organizations.

The Blue November movement is carried out by various organizations in November, aimed at raising awareness among men and society about the importance of prostate cancer prevention and early diagnosis, as well as other male health issues. "Movember" began in Australia in 2003, building on the celebration of World Prostate Cancer Awareness Day, held on November 17th.

Etymologically, the term prevention derives from the Latin *prevenire*, meaning "to come before," and it refers to the act of having an early opinion, so much so that "preventive" means something that is avoided. In its definition, prevention includes control actions aimed at avoiding the onset of specific diseases. In oncology, prevention is associated with primary and secondary levels.

Primary cancer prevention logically occurs before the disease arises, referring to any action aimed at reducing the population's exposure to risk factors, with the goal of decreasing the incidence of cancer through health promotion (non-specific measures, such as anti-smoking campaigns) and specific protection (such as vaccination).

Smoking. Responsible for about 90% of lung cancer deaths. Of all known cancer risk factors known to date, smoking, linked to cancers of the lung, upper airways, oesophagus, pancreas, bladder, and others, has been given special attention because it is also a disease caused by nicotine addiction. Smoking control is the main form of primary cancer prevention.

Obesity. According to a World Health Organization (WHO) report, up to one-third of colon, rectal, breast, and kidney cancer cases are attributed to excess weight and lack of physical activity. While smoking is a declining villain, obesity is a rising one. Obese individuals experience hormonal changes that contribute to the development of cancer cells.

Diet. Certain foods, when consumed regularly over long periods, seem to create conditions in parts of the body that are favourable to increased cell proliferation. This group includes high-fat foods such as red meats, fried foods, mayonnaise, dairy products, ham, bacon, and others, which should be avoided or consumed in moderation.

On the other hand, chemopreventive agents found in so-called functional foods can be used for a potent preventive effect, with multiple mechanisms of action that block carcinogenesis. In the case of breast cancer, for example, the preventive action of omega-3 polyunsaturated fatty acids, dietary fibres, and certain vitamins and minerals stands out.

Chemoprevention. Epidemiological studies, laboratory experiments, and many drugs used to treat established tumours provide chemopreventive agents. There are risk groups for this type of intervention, such as patients who have already had cancer, patients with premalignant lesions, and those at high risk for malignancy due to family history, lifestyle, or occupation.

In this context, raloxifene has been shown to prevent the development of new breast cancers as well as inhibit the growth of existing cancers. It has the same efficacy as tamoxifen in preventing in situ breast cancer, both reducing the risk by 50%. In fact, tamoxifen and raloxifene are already offered as preventive medications in some countries. Aspirin, which has been part of clinical practice for over 100 years, has also become particularly attractive as a potential agent in preventing colorectal and breast cancers.

Prophylactic surgery. Some congenital or genetic conditions pose a higher cancer risk to patients. These conditions, as well as the detection of precancerous lesions, may justify surgery to prevent cancer. For instance, a boy with cryptorchidism (when the testicle does not descend into the scrotum) should undergo orchidopexy (testicle fixation in the scrotum) to prevent testicular cancer.

Vaccination. In cases where cancer is linked to infectious agents, particularly oncogenic viruses, vaccine development can be highly beneficial. For example, the most important primary prevention of liver cancer today is vaccinating children and adults against hepatitis B. Liver cancer can be fully prevented if the main risk factors, such as chronic hepatitis B infections, are eliminated. However, the number of liver cancer cases will rise dramatically in the next 20 years if prevention strategies are not promoted.

Primary prevention of cervical cancer is associated with reducing the risk of HPV transmission, which occurs through sexual contact. Using condoms during penetrative sex offers partial protection against HPV transmission, which can also occur through skin contact with the vulva, perineal, perianal, and scrotal areas, or through the virus being deposited on fingers through genital contact and self-inoculation.

Additionally, there are intramuscular vaccines against HPV, which are most effective when administered before the onset of sexual activity, meaning to young individuals who have not yet been exposed to the virus. The first, bivalent (2vHPV), prevents infections and precancerous lesions of the cervix caused by HPV types 16 and 18. The quadrivalent vaccine (4vHPV), in addition to types 16 and 18, also prevents infections from types 6 and 11, which are responsible for 90% of genital warts and low-risk cervical lesions. Finally, the nonavalent vaccine (9vHPV) includes five additional HPV types, protecting against types 6, 11, 16, 18, 31, 33, 45, 52, and 58.

The implementation of the 9vHPV vaccine has the potential to prevent up to 93% of cervical cancer cases, as well as a significant proportion of other anogenital cancers related to HPV. The 9vHPV has been licensed by the U.S. Food and Drug Administration (FDA) since 2014 and is classified as a preventive immunization, routinely recommended for men and women aged 9 to 45 to prevent various HPV-related diseases and dysplastic lesions.

It is also worth mentioning genetic susceptibility testing. In cases of hereditary cancer syndromes, tests for potential mutations can be performed to identify family members at high risk of cancer. The decision to perform the genetic susceptibility testing should be the result of a consultation with a geneticist or oncologist specializing in cancer genetics. Upon a positive result, the physician can offer primary cancer prevention, such as through chemoprevention.

In cases of a strong family history of breast or ovarian cancer, for example, genetic testing can allow for the prevention of these malignancies. Testing for abnormalities in the *BRCA1* and *BRCA2* genes is now available, and a woman with a *BRCA1* or *BRCA2* mutation has a 60-85% risk of developing breast cancer and a 20-60% risk of developing ovarian cancer.

On May 14, 2013, actress Angelina Jolie revealed in *The New York Times* that she had undergone surgery to remove her breasts. Her decision to have a preventive double mastectomy was influenced by two strong factors: her family history (her mother died at 56 after nearly a decade battling cancer) and the discovery that she carried a defect in the *BRCA1* gene.

The actress stated that due to the *BRCA1* mutation, she faced an 87% risk of developing breast cancer and a 50% risk of ovarian cancer. Sometime later, on March 24, 2015, again through "*The New York Times*" in *The Opinion pages*, Jolie announced her new preventive surgery – this time involving the removal of her ovaries and fallopian tubes.

> It is a less complex surgery than a mastectomy, but its effects are more severe. It forces a woman into early menopause. [...] I went through what I imagine thousands of other women have felt. I told myself to stay calm, be strong, and that there was no reason to believe I wouldn't live to see my children grow up and meet my grandchildren (JOLIE, A. THE NEW YORK TIMES, 03/24/2015).

Secondary Cancer Prevention refers to the detection of pre-malignant or early-stage malignant lesions that are asymptomatic at a point where treatment can potentially be curative. For a screening test to be acceptable, it must meet certain criteria, such as high sensitivity and specificity, applicability to a large number of people, and ease and speed of administration.

Screening involves evaluating asymptomatic individuals to classify them as candidates for more refined diagnostic exams, with the aim of discovering hidden cancers or pre-malignant conditions that can be cured with treatment. Screening can include physical exams, self-examinations, radiological procedures, and laboratory tests. For example, breast cancer screening through mammography has demonstrated a significant overall reduction in mortality among women over 50 years old.

In the case of colorectal cancer, screening serves both primary prevention, by detecting precursor lesions (adenomatous polyps), and secondary prevention, by detecting early-stage malignant lesions. The simplest investigative test for detecting colon and rectal cancer is the faecal occult blood test, which identifies small amounts of blood in the stool.

A rectosigmoidoscopy is a procedure that allows direct visualization of the inner surface of the rectum and the lower third of the colon, the portion of the large intestine where most polyps and cancers occur. Another endoscopic examination is the colonoscopy, which enables the visualization of the entire colon, diagnosing problems in the colon and rectum, performing biopsies, and removing polyps.

The Pap test (Papanicolaou test) is the primary strategy used in cervical cancer screening programs. Women with epithelial cell abnormalities detected through the Pap test should undergo further testing with colposcopy and biopsy. A colposcope is a binocular microscope that allows for visual inspection of the cervix. Gross abnormalities observed during colposcopy can then be biopsied for further classification. It's important to note that HPV vaccines do not replace screening through preventive exams like the Pap test.

Cancer diagnosis

Cancer is one of the leading causes of mortality in our population. Many lives could be saved if cancer were diagnosed early. At this stage, certain changes can be observed that everyone should be aware of, as they serve as warning signs: a change in bowel or bladder habits; a sore that doesn't heal; unusual bleeding or discharge; a lump or thickening, such as in the breast or elsewhere; indigestion or difficulty swallowing; a noticeable change in a wart or mole; persistent cough and/or hoarseness.

The main reasons for late cancer diagnosis include the absence of symptoms in early lesions, lack of information about the disease, difficulty accessing healthcare, and the unpreparedness of healthcare professionals. On the other hand, the later the diagnosis, the lower the cure and survival rates, the higher the morbidity, and the greater the treatment costs.

The anamnesis (an interview conducted by the healthcare professional with the patient) and the physical exam form the foundation of clinical diagnosis and guide the need for additional tests. These tests in oncology are requested to assess the primary tumour, organ functions, the presence of other diseases, and the extent of the neoplastic disease (staging).

The diagnosis of cancer should include a detailed medical history and physical examination. Whenever possible, it should also involve direct visualization of the affected area through endoscopic procedures such as bronchoscopy, upper gastrointestinal endoscopy, mediastinoscopy, pleuroscopy, rectosigmoidoscopy, colonoscopy, urological endoscopy, laryngoscopy, and colposcopy.

If any abnormalities are observed, tissue samples should be biopsied and sent for histopathological examination to confirm or rule out the diagnosis. A small tissue sample is surgically removed and examined under a microscope by a pathologist to detect the presence of neoplastic cells. Some biopsies can be performed in an outpatient setting using only local anaesthesia.

Additional tests are also recommended for monitoring treatment, detecting recurrences, and screening high-risk groups. These tests, often the same as those used to diagnose other diseases, provide anatomical and functional assessments of the patient, evaluation of the primary tumour, and identification of local, regional, or distant complications.

Some tumours are characterized by the production of certain substances, which can be measured for diagnostic purposes, staging parameters, treatment monitoring, and prognostic factor. These substances are known as tumour markers and, depending on their levels, results can be interpreted as normal, suspicious, or pathological.

In addition to enabling early diagnosis, tumour markers can help identify the affected organs and assess the extent of the disease. Many of these markers already have recognized clinical value. Notable examples include carcinoembryonic antigen (CEA), associated with gastrointestinal tract tumours; beta-2 microglobulin (β2M), linked to multiple myeloma and malignant lymphoma; and prostate-specific antigen (PSA), related to prostate carcinoma.

The diagnosis of prostate cancer is confirmed through the histopathological study of prostate tissue obtained via biopsy, which should be considered whenever abnormalities are found during a digital rectal exam or elevated PSA levels are detected. The anatomopathological report should provide histological grading, which helps determine the tumour's likely growth rate and its potential for spreading.

Cancer diagnosis can also be performed using cytological techniques, which involve the microscopic examination of stained cells. There are various methods for collecting and staining cell samples. One approach uses a fine needle and syringe to aspirate cells from palpable surface masses; another is exfoliative cytology, commonly used to obtain cervical smears.

The study of chromosomal abnormalities in neoplastic cells is highly valuable for diagnosing, classifying, guiding treatment, and predicting the prognosis of leukaemia. Numerical and structural karyotypic abnormalities are confined to malignant clones, disappear during hematologic remission, and reappear with relapse, sometimes showing new alterations.

Cancer treatment

Different types of cancer have distinct patterns of neoplastic cell spread. For example, breast cancer tends to metastasize to the bones, lungs, liver, and brain. Each type of cancer also has specific tests recommended for diagnosis and treatment based on the stage of the disease. In turn, the primary goals of treatment are cure and control.

A cure is commonly defined when no further treatment is needed, or when there is no evidence of cancer recurrence for five years. Treatment may also aim to control cancer that has recurred locally or metastasized. In these cases, the objective is to manage the cancer's growth or spread for as long as possible.

When well-known public figures are diagnosed with cancer, media coverage often brings the disease closer to the general public, even for those not directly affected. For instance, Australian actor Hugh Jackman, famous for his role as Wolverine, had several skin cancers surgically removed from his face between 2017 and 2019. On the other hand, Brazilian soccer legend Edson Arantes do Nascimento, known as Pelé, widely regarded as the greatest footballer of all time, passed away in December 2022 after a prolonged battle with colorectal cancer.

There are four main therapeutic modalities for cancer: surgery, radiotherapy, chemotherapy, and biological therapy. Treatment often involves using more than one of these modalities in combination. The goal is to achieve higher cure rates while minimizing anatomical loss, preserving aesthetics and the function of affected organs, and reducing toxicity.

Oncological Surgery. Surgical procedures are the treatment of choice for tumours confined to a single location. Surgery is used to remove a defined mass believed to be malignant. It is the preferred option for the complete removal of solid tumours (such as breast, colon, and lung cancers), along with a margin around them if the cancer is locally confined.

Surgery can be employed as a standalone treatment or in combination with other modalities, such as radiotherapy, chemotherapy, or biological therapy. Additionally, surgery has various indications: it can be used for diagnosis, staging, prevention, and treatment; for reconstruction, inserting therapeutic and supportive devices; and for addressing cancer-related complications.

Radiotherapy. This therapeutic modality uses ionizing radiation (such as X-rays) or high-energy particles. According to the World Health Organization (WHO), approximately 70% of cancer patients will undergo radiotherapy at some point during their treatment. Ionizing radiation disrupts the atomic structure of molecules, leading to the death of affected cells. This is why it is used to treat tumours that selectively absorb radiation, meaning the tumour cells absorb more radiation than the healthy surrounding tissues.

Radiotherapy halts malignant growth by causing two types of damage once ionizing radiation is absorbed by living tissue. The first is direct damage, such as lesions to the DNA molecule at the chromosomal level. The second is indirect damage, resulting from a series of reactions that produce free radicals, which can have cytotoxic effects.

Rapidly reproducing cells, such as cancer cells, are more sensitive to radiation since it primarily affects cells during mitosis, although damage can occur in other phases of the cell cycle as well. Well-oxygenated cells are also highly radiosensitive. The side effects of radiotherapy result from damage to normal cells within the treatment field and may include skin issues, mucositis, nausea and vomiting, and loss of appetite.

Chemotherapy. Also known as cytotoxic therapy, chemotherapy involves the use of drugs aimed at cancer cells, administered either during hospitalization or on an outpatient basis. Unlike surgery and radiotherapy, which are local treatments, the rationale behind drug therapy for cancer is that it is a systemic treatment capable of reaching cancer cells even if they have already spread throughout the body.

While chemotherapy can consist of a single drug, treatment regimens typically combine different drugs. This approach allows malignant cells to be attacked by drugs with various mechanisms of action, thereby enhancing the treatment's effectiveness and reducing the resistance of cancer cells. These mechanisms include antimitotic agents (e.g., paclitaxel), topoisomerase inhibitors (e.g., camptothecin), and antineoplastic antibiotics (e.g., doxorubicin).

Targeting cancer with drugs that inhibit cell multiplication, one of the key characteristics of neoplasms, also affects healthy cells that rapidly reproduce, such as those responsible for hair growth, the body's immune defence, and cells in the digestive tract. This explains why the main side effects of chemotherapy are linked to the functioning of these cells.

The most common side effects fall into four categories: alopecia, which refers to hair loss, particularly of the scalp; general immune suppression, increasing the likelihood of opportunistic infections; neurotoxicity, often felt as tingling in the hands or feet; and gastrointestinal issues such as diarrhoea, nausea, vomiting, and loss of appetite.

Biological Therapy. Sometimes referred to as biotherapy, this is a relatively new approach in the treatment of various conditions, such as autoimmune diseases, inflammatory disorders, and cancer. Generally, these are immunotherapies that, in the case of cancer, utilize the body's immune system, either directly or indirectly, to fight the tumour or to reduce the side effects that may arise from certain antitumor treatments.

Monoclonal antibodies are laboratory-produced versions of immune system proteins designed to target a specific molecule. In breast cancer, for example, the tumour can be tested for the expression of a protein encoded by the *HER-2* oncogene. If the gene expression is increased (high protein levels), the tumour becomes more aggressive and less responsive to chemotherapy. However, there is a biological therapy available – an antibody against the HER-2 protein (anti-HER2 monoclonal antibody), which can reverse this situation.

Blinatumomab is a special type of monoclonal antibody developed to treat patients with B-lineage Acute Lymphoblastic Leukaemia (ALL). One part of the drug binds to the CD19 protein found on B cells, including some leukaemia and lymphoma cells. The other part binds to the CD3 protein found on T cells of the immune system. By binding to both proteins, blinatumomab prompts the immune system to attack cancer cells.

Cytokines are another important group of proteins in this field, playing a crucial role in communication between immune cells. In this type of approach, immune cells are taken from the patient's blood and treated with cytokines in a laboratory to activate them. These cells are then reinjected into the patient, with the hope of boosting the immune system's ability to fight the disease.

Among cytokines, interleukins and interferons stand out for their role in coordinating the immune response. Interleukins are various glycosylated or non-glycosylated proteins, mainly produced by white blood cells, which act in inflammatory processes and immune responses. For example, infusions of interleukin-2 (IL-2) have led to complete and lasting regressions of metastases in cases of melanoma and kidney cancer.

Interferons are glycoproteins originally discovered as molecules that interfere with viral replication during infection. Produced by leukocytes and fibroblasts, they can inhibit the proliferation of viruses, bacteria, fungi, and cancer cells. Interferon-alpha, for instance, has been approved for use against melanoma, certain types of leukaemia, and Kaposi's sarcoma. As with other cancer treatments, side effects can occur, including fever, fatigue, chills, muscle pain, headaches, anorexia, and nausea.

In addition to biopharmaceuticals, oncology biotechnology also focuses on natural and artificial microRNAs (miRNAs), which present a promising avenue for cancer therapy. MicroRNAs are small non-coding RNA molecules that regulate gene expression by inducing degradation or repressing the translation of target messenger RNA molecules. Several miRNAs have shown high efficacy in both *in vitro* and *in vivo* studies. However, no miRNA has yet been approved by the U.S. FDA, and many have been discarded due to toxicity issues. These setbacks highlight the challenges that must be addressed for the widespread application of miRNA-based therapies.

Gene therapy is an alternative treatment method that introduces healthy (therapeutic or target) genes into cancerous cells or tissues. This approach may involve strategies such as activating therapeutic genes or silencing undesirable or defective ones, aiming to either kill or slow the growth of cancer cells. There are two main techniques to deliver genetic material into a patient's cells: *ex vivo*, where cells are isolated, modified in the lab, and then returned to the patient's body; and *in vivo*, where the genes are inserted directly into the cells using a vector.

For gene introduction into cells, a vehicle called a vector is needed, which facilitates the entry of DNA into the targeted cells. These vectors fall into three main categories: viral vectors, which include either harmless viruses or harmful ones that are engineered to lose undesirable traits; plasmids, small circular DNA segments; and nanostructured vectors, polymers that form networks to trap a gene and release it once inside the cells.

An example of gene therapy already approved in several countries is CAR-T Cell Therapy, which uses a patient’s own cells to target specific cancer cells. T-cells are taken from the patient and engineered in the lab by introducing a gene for CAR (Chimeric Antigen Receptor), enabling them to fight the cancer. These modified T-cells (CAR-T Cells) can find and kill cancer cells that have a specific protein on their surface, which the receptor is designed to bind to. This technology has been particularly effective in treating leukaemias and lymphomas.

Recent advances are steering cancer therapies toward increasingly personalized approaches. For example, current CAR-T Cell therapies are tailored to each individual patient. In addition to differences between individuals, there is a wide variety of genetic profiles in neoplasms. The better characterized a patient's cancer cells are, the greater the chances of success in selecting the most effective treatments and drugs for a specific type of tumour. Personalized medicine aims to enhance treatment efficacy and plays a vital role in early cancer diagnosis and prevention.

Among the most recent studies in the search for new cancer therapies is the publication "Paradoxical activation of oncogenic signalling as a cancer treatment strategy" by Matheus Henrique Dias and collaborators. This innovative approach involves intentionally hyperactivating oncogenic signalling combined with disrupting stress responses. Promising results were achieved in the treatment of colon cancer in animal models. This approach is expected to be tested on patients with intestinal tumours in the Netherlands in 2025.

READING SUGGESTIONS

An Approach to the Primary and Secondary Prevention of Gastric Cancer in the United States
Clinical Gastroenterology and Hepatology
Robert J. Huang, Meira Epplein, Chisato Hamashima, and collaborators.

Influence of the Angelina Jolie Announcement and Insurance Reimbursement on Practice Patterns for Hereditary Breast Cancer
Journal of Breast Cancer
Jihyoun Lee, Sungwon Kim, Eunyoung Kang, and collaborators.

Paclitaxel
StatPearls [Internet]
Ayoola O. Awosika, Maela C. Farrar, and Tibb F. Jacobs.

Monoclonal Antibodies in Cancer Therapy
Antibodies (Basel, Switzerland)
David Zahavi, and Louis Weiner.

10. NATURAL PRODUCTS WITH ANTITUMOR ACTIVITY

Introduction

Natural products, or secondary metabolites, are substances found in a single type of organism or in a small group of genetically related organisms. These natural products act by mediating specific ecological interactions between organisms and their environment, as well as interactions among organisms of the same species or between different species.

In evolutionary and ecological studies, these substances are typically referred to as semiochemicals, which are further classified into pheromones and allelochemicals. Pheromones mediate interactions between organisms of the same species, while allelochemicals mediate interspecies interactions. When allelochemicals benefit only the producing species, they are called allomones.

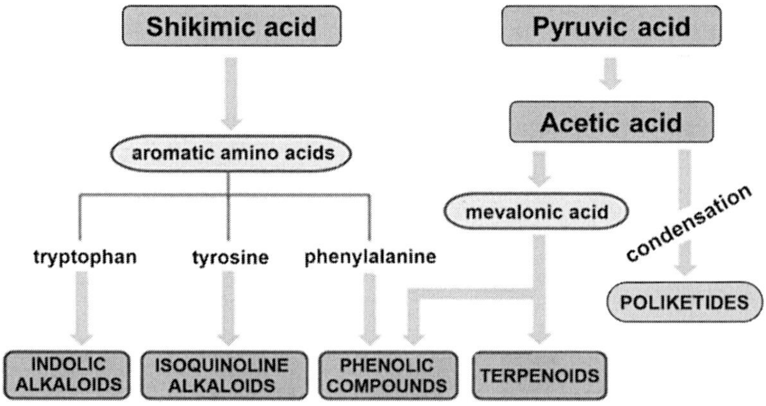

Figure 10-1. Secondary Metabolites. The main pathways of secondary metabolism and their interconnections.

Secondary metabolites are often composed of complex chemical structures with well-defined spatial orientation. They belong to major groups such as alkaloids, phenolic compounds, terpenoids, and polyketides. As shown in Figure 10-1, they derive from two main biosynthetic pathways: (a) the shikimic acid pathway and (b) the pyruvic and acetic acid pathway.

Natural products or specialized metabolites have been sources of food, fragrances, pigments, insecticides, and medicines. Their significance is such that over 25% of commercial drugs are derived from natural plant products, and at least 12% come from microorganisms. Since ancient times, humans have relied on elements from nature as sources of medicine.

Our search for substances in living organisms for therapeutic purposes is based on the vast chemical diversity found in species of plants, animals, fungi, and microorganisms, which ultimately stems from evolution. This diversity reflects the selection and conservation of key defence mechanisms used by organisms to paralyze, repel, or kill predators.

The discovery or development of effective drugs is a complex, costly, and highly coordinated process. Obtaining satisfactory results requires teamwork and experts from various fields: Biological and Health Sciences, Earth and Exact Sciences, Engineering. It is a multidisciplinary activity that has had Medicinal Chemistry as its central axis since its beginnings.

Anticancer drugs

Leticia Veras Costa-Lotufo from the University of São Paulo (Brazil) states that the development of anticancer drugs has historically been linked to natural products, with over 60% of the drugs currently used having at least a partial origin from natural sources. On the other hand, while new drugs are continuously being added to the therapeutic arsenal against cancer, there are still numerous tumours for which we lack adequate treatment.

Many natural compounds have been identified through ethnopharmacological knowledge of certain plants and animals, with herbal medicine and zootherapy often serving as starting points for current research. The table below provides a summary of the therapeutic indications of some natural drugs and their derivatives currently used in cancer chemotherapy.

Table 10.1. Natural products or derived drugs used in cancer therapy.

Major groups		**Species**	**Drugs**	**Therapeutic indications**
Plantae	Gymnospermae	*Taxus brevifolia*	Paclitaxel	Breast, ovarian, and lung cancer.
	Angiospermae	*Camptotheca roseus*	Camptothecin	Colorectal cancer.
		Catharanthus acuminata	Vincristine	Acute lymphoblastic leukaemia; Hodgkin's lymphoma; osteogenic sarcomas.
Animalia	Porifera	*Halichondria okadai*	Halichondrin B	Breast cancer.
	Mollusca	*Dolabella auricularia*	Dolastatins	Melanomas.
		Elysia rufescens	Kahalalide F	Melanomas; hepatocellular carcinoma.
	Urochordata	*Ecteinascidia turbinata*	Ecteinascidin-743	Soft tissue sarcoma.
	Chondrychthyes	*Squalus acanthias*	Neovastat	Renal carcinoma, multiple myeloma.

As expected, considering the versatility of pathways that can be involved in the carcinogenesis process, the mechanisms of action of antineoplastic drugs are also quite diverse. Most antineoplastic agents belong to the group of cell cycle-specific drugs, acting on cells in the process of division, but there are also drugs that act independently of the cell cycle. Among the agents that act dependent on cell proliferation, some have preferential action in a specific phase of the cycle, known as phase-specific drugs, while others are cytotoxic at any point in the cycle – phase-non-specific drugs. To illustrate, here are some examples of drugs and their mechanisms of action.

Camptothecin is a cytotoxic alkaloid with considerable antitumor activity. Its mechanism of action involves inhibiting the enzyme DNA topoisomerase, which is often present in high concentrations in tumours. This action is significant because topoisomerase plays a critical role in DNA replication, transcription, recombination, and chromatin remodelling.

Vincristine is another alkaloid and acts as an antimitotic. Its mechanism of action involves the depolymerization of microtubules, which are an important part of the cytoskeleton and the mitotic spindle. The drug specifically binds to tubulin, disrupting its organization, leading to the depolymerization of microtubules, the dissolution of the mitotic spindle, and halting mitosis at metaphase.

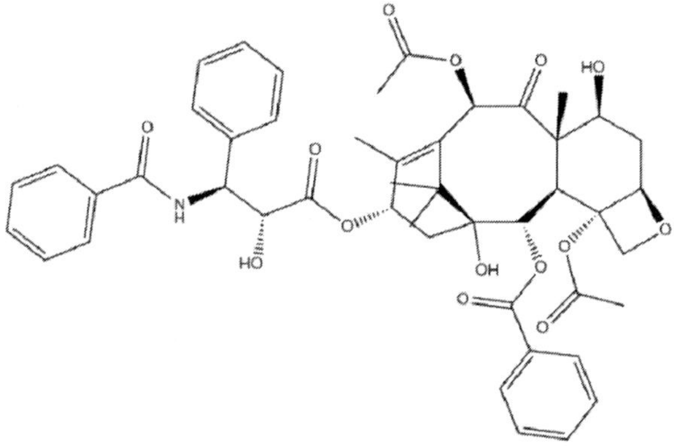

Figure 10-2. Structure of paclitaxel. The molecule consists of a tetracyclic core called baccatin III and an amide tail. Its molecular formula is $C_{47}H_{51}NO_{14}$, and it can be identified as an ester of baccatin III N-benzoyl-β-phenylisoserine.

Paclitaxel (commercially known as Taxol®) is yet another antimitotic alkaloid that interferes with microtubule function, preventing uncontrolled cell division. However, it acts differently: the tubulin/paclitaxel complex results in the formation of very stable and non-functional microtubules.

This means that the cell is unable to use its cytoskeleton in a flexible manner. New studies have also indicated that paclitaxel can reactivate the immune response, thus fighting cancer on two fronts.

Kahalalide F is a depsipeptide whose mechanism of action disrupts the basal function of lysosomal membranes, inducing cell death. It's worth explaining here: the term "depsipeptide" refers to a peptide in which one or more amide bonds (-CONHR-) are replaced by ester bonds (-COOR-). In Figure 3, we can see a simulation of the production of a depsipeptide.

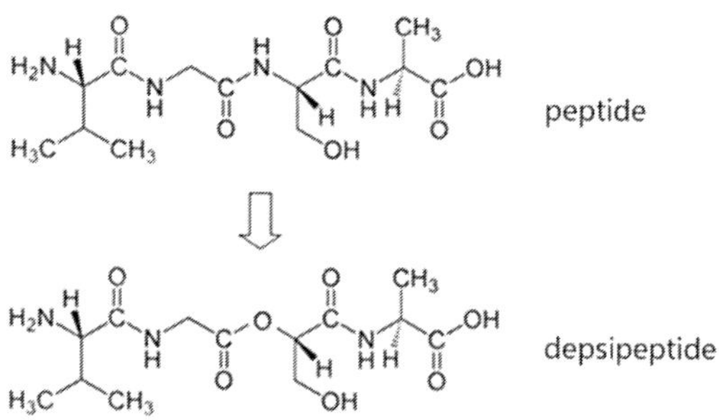

Figure 10-3. Production of a depsipeptide. At the top, the structure of a peptide (specifically a tetrapeptide) with three peptide bonds (amide bonds). In the simulation, the second amide bond has been replaced by an ester bond, resulting in a depsipeptide.

Looking at the drugs summarized in Table 10-1, we can see that, in addition to classic terrestrial sources like plants and soil, research on marine natural products has also gained prominence. With advancements in safe diving techniques and equipment, algae and invertebrates have started making their way into the chemistry and pharmacology labs.

> The achievements and progress in marine natural product research have already earned their rightful place in the pharmaceutical industry, and the promise of marine-derived drugs has undeniably become a reality. [...] In this context, the chemical diversity of natural products from marine organisms, combined with the plurality of therapeutic targets, offers innovative perspectives for the clinical approach to malignant neoplasms (COSTA-LOTUFO et al., 2010, p. 713-714).

Over the past 30 years, marine organisms of different taxonomic groups have become highly intriguing sources for isolating bioactive natural products. Sponges, cnidarians, molluscs, and ascidians, along with algae, are among the most chemically prolific marine groups, producing a wide variety of structurally interesting and biologically active compounds.

In selecting organisms for research, soft-bodied, sessile animals, often lacking obvious physical defence mechanisms and frequently brightly coloured, deserve priority. We can imagine that, through natural selection, they would have become easy prey, leading to their depletion or extinction. Since that is not the case, they are strong candidates for possessing chemical defences.

Indeed, the cytotoxicity of marine natural products is the most commonly sought biological activity and typically exhibits great potency, which aligns with the proposed ecological role of these compounds: to aid in competition for space, eliminate pathogenic microorganisms, and paralyze or kill predators, overcoming the diluting effects of the aquatic environment.

Biotechnological potential of microorganisms

According to Ricardo Pereira from the Fluminense Federal University, Rio de Janeiro (Brazil), the ability to synthesize a wide diversity of metabolites through such distinct biosynthetic pathways is not a characteristic of animals, but rather an activity specific to microorganisms associated with them. These metabolites are almost exclusively produced by the associated flora or obtained from dietary sources.

The ascidian *Cystodytes dellechiajei* inhabits benthic rocky environments in tropical and temperate waters of the Atlantic, Pacific, and Indian Oceans, as well as the Mediterranean Sea. The alkaloids known as pyridoacridines are the main compounds extracted from this ascidian. In 2007, Manuel Martínez-García and collaborators from the University of Alicante (Spain) published a description of the *in vitro* antiproliferative activity of *C. dellechiajei* extracts against different tumour cell lines.

The extracts from *C. dellechiajei* showed remarkably high cytotoxic activity against the A-549 lung carcinoma, H-116 colorectal adenocarcinoma,

PSN-1 pancreatic adenocarcinoma, and SKBR3 breast carcinoma cell lines. Additionally, this study also provided insights into the role of the microbial community associated with the invertebrate in the production of the active compounds.

> Natural products are a fundamental source of new chemical diversity and an integral component of the current pharmaceutical compendium. [...] While marine plants and invertebrates have received considerable attention as a resource for natural product discovery, the microbiological component of this diversity remains relatively unexplored. Recent studies have revealed that selected groups of marine actinomycetes are a robust source of new natural products (FENICAL & JENSEN, 2006, p. 666).

More recently, researchers in natural products have turned their attention to microorganisms, driven motivated both by the prolificacy of these organisms and by the need to address a major issue in obtaining drugs from animals in the wild: due to the very low yield, large amounts of biomass would need to be harvested, leading to significant environmental harm.

The ascidian *Ecteinascidia turbinata*, commonly found in shallow tropical and subtropical habitats, has been recorded in both the eastern and western Atlantic Ocean, as well as the Mediterranean Sea. Trabectedin (Ecteinascidin-743; Yondelis®), isolated from this ascidian, was the first antitumor drug to gain clinical approval for treating soft tissue sarcoma. The European Medicines Agency (EMEA) granted marketing authorization for the European Union in September 2007.

Previously, *E. turbinata* had already been successfully cultivated, and for clinical trials, E. J. Corey and collaborators developed a synthesis method in 1996, which was licensed to PharmaMar. However, industrial application proved impossible due to low yields. Currently, its production is carried out through a semisynthetic process starting from cyanosafracin B, which is bulk-produced via fermentation of the marine bacterium *Pseudomonas fluorescens*.

During my postdoctoral research (2011-2012), I worked on bioprospecting antitumor substances obtained from microorganisms associated with the cnidarians *Palythoa caribaeorum* and *Protopalythoa variabilis*, found along the coast of Ceará, Brazil, in the National Marine Biodiversity Research Network (SISBIOTA-MAR). By the end of the isolation and purification processes, 20 bacterial strains were obtained, and their extracts were tested against the PC-3/M (metastatic prostate cancer), OVCAR-8 (ovarian cancer), and HCT-116 (colorectal adenocarcinoma) cell lines. The strain identified as BRA-035, a Gram-positive isolated from *P. variabilis*, showed the highest activity.

The BRA-035 strain then received further attention, including its molecular characterization through partial sequencing of the 16S rRNA gene. The sequence obtained (about 1400 bp) was compared using the National Center for Biotechnology Information (NCBI) database, employing the BLAST tool. Molecular identification through 16S rRNA sequencing revealed that BRA-035 belongs to the *Streptomyces* genus. Moreover, bio-guided fractionation of the BRA-035 extract led to the isolation of substances like piericidin A, the primary agent responsible for the high antiproliferative activity observed in the exposed cell lines.

Another interesting example is salinosporamide A, which is still in clinical trials for multiple myeloma and other types of cancer. It is extracted from the bacterium *Salinispora tropica*, collected from ocean sediments in the Bahamas. This compound holds the distinction of being the substance that moved most quickly through preclinical trials, taking only three years from its discovery to the start of clinical testing.

The search for new cancer-fighting molecules also includes research on terrestrial microorganisms. For instance, temsirolimus, commercially known as Torisel®, was approved by the U.S. FDA in May 2007 for the treatment of advanced renal cell carcinoma. The precursor to temsirolimus is the antibiotic rapamycin, which was isolated from the soil bacterium *Streptomyces hygroscopicus*.

Current research

Since an ideal compound for the treatment of malignant tumours has yet to be discovered, the search continues for more selective compounds with higher therapeutic potency, fewer side effects, and lower resistance rates. Natural resources, as sources of secondary metabolites, remain abundant and offer a higher probability of finding therapeutically valuable substances. Over the years, strategies for discovering new drugs have evolved. Experimental Oncology aims to study the mechanisms of tumour development in experimental models and explore potential treatment options.

Regarding microorganisms, the research typically begins with their isolation and cultivation in culture media. Afterward, secondary metabolites are extracted using appropriate solvents, and the extracts are tested *in vitro* against cancer cell lines. If an extract demonstrates significant cytotoxicity, inhibiting the proliferation of cancer cells, the process moves on to bio-guided fractionation.

In this phase, each crude extract, which contains complex mixtures, undergoes fractionation and purification. It is termed "bio-guided" because cytotoxic activity testing is repeated at each stage, from fractions to pure substances. Once a pure compound is obtained, further testing can be conducted to assess its cytotoxicity against a broader panel of cell lines, its selectivity, and its mechanism of action.

Following *in vitro* studies, tests on non-human animal models are performed to confirm *in vivo* activity, which requires a larger quantity of material. The first models typically used are mice and rats, sometimes in xenograft models, that is, genetically modified immunodeficient animals that are implanted with human tumours in their subcutaneous tissue.

The preclinical stage involves testing on animal models (including primates) to assess the viability, safety, and efficacy of compounds that could potentially become drugs. The clinical stage, which involves human trials, is typically divided into three phases. This entire process is lengthy, taking around 15 years from the discovery of a molecule to its availability on pharmacy shelves.

While most chemotherapy drugs used against cancerous tumours have been selected for their antiproliferative activity, recent research is also targeting new therapeutic avenues. Drugs have been identified that specifically act on certain points of tumour cell metabolism, inhibit tumour neovascularization, induce cell redifferentiation, or trigger apoptosis.

Thanks to recombinant DNA technology, it is now possible to alter the genetic material of an organism to achieve desired traits and products, such as proteins. Gene cloning and the expression of heterologous proteins in bacteria have become viable alternatives for synthesizing secondary metabolites of interest, including certain polyketides. Additionally, synthesis methodologies have advanced, allowing for the production of sufficient quantities of molecules for clinical studies and eventual commercialization.

Heterologous proteins are so named because they are produced by cells that do not naturally generate them. In this process, genes essential for producing a bioactive molecule are extracted from an animal, plant, or microorganism and inserted into specific bacteria for amplification and expression. The species *Escherichia coli* stands out among the bacteria most commonly used for this purpose.

READING SUGGESTIONS

The Role of Bioactive Compounds in Natural ProductsExtracted from Plants in Cancer Treatment and Their Mechanisms Related to Anticancer Effects
Oxidative Medicine and Cellular Longevity
Meng Yuan, Guoqing Zhang, Weijun Bai, and collaborators.

Recent Advances in Small Peptides of Marine Origin in Cancer Therapy
Marine drugs
Qi-Ting Zhang, Ze-Dong Liu, Ze Wang, and collaborators.

GLOSSARY

Adenocarcinoma. A carcinoma of secretory epithelial cells, such as gastric adenocarcinoma. See also carcinoma.

Aetiology. The study of causes. In medicine, for instance, it investigates the causes of diseases.

Allele. One of two or more alternative forms of a gene. The alleles of a gene are very similar in their nitrogenous base sequence, but each codes for a slightly different protein. In some cases, an allele is inactive, meaning it has lost the ability to code for a protein.

Aneugenic agents. Chemical or physical agents that induce alterations in the number of chromosomes in cells.

Apoptosis. ***See*** **programmed cell death.**

Atrophic gastritis. A histopathological condition characterized by chronic inflammation of the gastric mucosa, with loss of gastric glandular cells and their replacement by intestinal-type epithelium and fibrous tissue. It represents the endpoint of chronic processes such as chronic gastritis associated with *Helicobacter pylori*, other unidentified environmental factors, and autoimmunity targeting the stomach's glandular cells.

Bioassay. A controlled experiment aimed at estimating the effect of a particular substance on a living organism. Also known as a biological assay.

Blasts. Refers to young cells. Examples: erythroblasts, osteoblasts.

Carcinoma. A cancer of epithelial tissues, such as squamous cell gastric carcinoma, colon carcinoma, and skin carcinoma. ***See also*** **adenocarcinoma.**

Caspases (*cysteine-dependent aspartate-specific proteases*). Enzymes belonging to the cysteine protease family (with a cysteine in their active site) that can recognize and cleave substrates containing aspartate residues. Caspases signal apoptosis and cleave these substrates, leading to nuclear condensation and fragmentation, as well as externalization of membrane phospholipids, which signal macrophages to phagocytose these cells.

CDK (*Cyclin-dependent kinase*). A protein kinase that must complex with a cyclin to function. Different cyclin-CDK complexes activate various stages of the cell cycle by specifically phosphorylating target proteins.

Cell cycle. A sequentially ordered series of events that ensures a somatic cell duplicates its genetic material and divides it equally between its daughter cells.

Cell differentiation. The process by which a cell undergoes changes to become a truly specialized cell for a specific function or set of functions.

Cellular signalling: The process by which cells are stimulated or inhibited by extracellular signals, usually chemicals secreted by other cells.

Central dogma. The principal hypothesis in molecular genetics, proposed by Francis Crick in 1958, which states that DNA forms RNA, which then forms a protein.

Centrosome. An organelle found in the somatic cells of animals, centrally located, which serves as the primary microtubule-organizing centre and acts as a spindle pole during mitosis. The centrosome in most animals contains a pair of centrioles at its core.

Checkpoints. Key points in the cell cycle where the cell evaluates both external and internal conditions and decides whether to proceed to the next phase, preventing subsequent events from starting until the previous ones have been successfully completed. Internal checks occur to ensure the cell maintains the integrity of the genetic material being duplicated and passed on:

- **G1/S checkpoint:** At the end of G1, before DNA synthesis begins, it relies on mechanisms that monitor DNA integrity.

- **G2/M checkpoint:** After DNA duplication, it ensures the complete and accurate replication of the genome.
- **M checkpoint:** Monitors the attachment of chromosomes to the mitotic spindle.

Chemoprevention: The use of natural or synthetic chemical agents (drugs) to prevent the onset of cancer or block an initial stage of carcinogenesis.

CKIs (*Cyclin-dependent kinase inhibitors*). Proteins that interact with CDKs or cyclin-CDK complexes, blocking their kinase activity. In this state, the kinases no longer phosphorylate proteins, leading to cell cycle arrest.

Clastogenic agents. Chemical or physical agents that cause chromosomal breaks.

Cyclin. A protein whose concentration periodically rises and falls according to the eukaryotic cell cycle. Cyclins activate essential protein kinases (CDKs) and therefore help regulate the progression from one stage of the cell cycle to the next.

Cytokines. Proteins that modulate the function of other cells or the cell that produced them. They are produced by various cells, especially lymphocytes and activated macrophages, and are crucial in controlling the immune response.

Deamination. A change in the nitrogenous bases of DNA characterized by the replacement of an amine group (-NH2) with a hydroxyl group (-OH). Like tautomerization, deaminated bases behave as unusual bases and cause incorrect pairings.

***de novo* mutation.** A new mutation, meaning one that occurs in isolated cases, not linked to family history. This type of mutation introduces a sporadic trait to the occurrence of a genetic disease.

E_2F Protein: A regulatory protein that activates several genes encoding proteins needed for the cell to enter the S phase of the cell cycle.

Endoreduplication. A process where the nucleus undergoes cycles of DNA synthesis without cytokinesis, resulting in polyploid cells.

Ependymoma. A tumour originating from the ependyma, or neuroepithelium, that lines the brain's ventricles and the spinal canal.

Erythema. A reddening of the skin caused by local capillary vasodilation. It is a clinical sign that can be present in various pathologies.

Erythemogenic. Refers to the property or potential to cause erythema.

Exon. Each of the coding regions of a discontinuous gene.

General transcription factor. Any protein that helps position eukaryotic RNA polymerase correctly on the promoter, aids in separating the two DNA strands to allow transcription to begin, and releases RNA polymerase from the promoter into elongation mode once transcription has started.

Growth factors. Protein substances involved in cellular signalling. They stimulate cell proliferation and may also affect cell contractility, motility, differentiation, and angiogenesis.

Haematological parameters. Parameters assessed in a blood test. Examples include red blood cells (RBC), haemoglobin (HB), haematocrit (HT), mean corpuscular volume (MCV), mean corpuscular haemoglobin (MCH), and mean corpuscular haemoglobin concentration (MCHC).

Heterochromatin. The more condensed portion of chromatin within the cell nucleus, with low gene activity, in contrast to euchromatin, which is less condensed and exhibits high transcriptional activity.

Histiocytes. Cells derived from monocytes that act as fixed macrophages in connective tissue.

Hyperplasia. The increase in size of a tissue or organ caused by an increase in the number of cells.

Hypertrophy. The increase in the size of a tissue or organ due to changes in cell volume, resulting from enhanced cellular functions.

Integrins. Proteins involved in cell adhesion or anchoring to the extracellular matrix, also playing a role in cell proliferation and differentiation.

Intron. A segment of a discontinuous gene that does not contain biological information.

***In vitro*.** A biological process or reaction occurring outside the body under artificial conditions, typically in a test tube or culture plate.

***In vivo*.** A biological process or reaction occurring within a living cell or organism.

Isotopes. Atoms with the same number of protons but different mass numbers. For example, the three isotopes of hydrogen have special names: hydrogen, deuterium, and tritium. Isotopy is a common natural phenomenon, as nearly all natural chemical elements are composed of isotope mixtures.

Kinase: An enzyme that transfers the terminal phosphate group from ATP to one or more specific amino acids (serine, threonine, or tyrosine) in a target protein. Also known as a protein kinase.

Kinetochore. A complex protein structure formed at the centromere of a metaphase chromosome, where microtubules attach. It plays an active role in moving chromosomes toward the poles.

Leukocytes. White blood cells, including various types such as lymphocytes, neutrophils, and monocytes.

Mast cells. Connective tissue cells that participate in inflammatory responses by releasing various molecules accumulated in their cytoplasm, such as histamine. Mast cells are crucial in allergic reactions.

Melanogenesis. The process by which melanin is produced within melanocytes, cells found in the basal layer of the epidermis.

Mesenchymal cells. Elongated cells with oval nuclei, formed during differentiation of the middle embryonic layer (mesoderm) and functioning as stem cells.

Messenger RNA (mRNA): An RNA molecule that functions during translation to specify the amino acid sequence of a newly formed polypeptide.

Metaphase chromosome. During metaphase, the chromosome consists of two sister chromatids held together by a narrow region called the primary constriction or centromere. Transversely, the centromere divides the chromatid into two chromosome arms. The long arm is labeled "q," while the short arm is labelled "p" (from the French word *petit*).

Mitogens. Compounds that induce mitotic divisions, leading to cell proliferation.

Morbidity. The number of cases of a particular disease relative to the total population being studied.

Mucositis. Inflammation and ulceration of the mucous membranes lining the digestive tract. In patients undergoing chemotherapy and/or head and neck radiation therapy, oral mucositis tends to be exacerbated.

Mutations. Variations that occur and become fixed in genetic material. They can involve anything from small sections of a DNA molecule to entire chromosomes.

Natural selection: An evolutionary mechanism that causes changes in the frequency of specific genotypes in a population over generations, due to differential survival and/or reproduction.

Necrosis. A type of cell death where cells suffer an insult resulting in increased cell volume, chromatin aggregation, cytoplasmic disorganization, loss of plasma membrane integrity, and eventual cell rupture. During necrosis, cell contents are released, causing damage to neighbouring cells and triggering an inflammatory response in the area.

Negative regulators of the cell cycle. Enzymes or protein complexes that inactivate the functions of positive regulators, causing cell cycle arrest and apoptosis (programmed cell death). There are three main types: cyclin-dependent kinase inhibitors, ubiquitin complexes, and phosphatases.

Nucleic acid probe: A defined fragment of RNA or DNA, chemically or radioactively labelled, used to locate specific nucleic acid sequences through hybridization.

Nucleoside. A compound formed by a nitrogenous base covalently bonded to a pentose (ribose or deoxyribose); for example, adenosine and guanosine.

Nucleotide. The building block of nucleic acid molecules, consisting of a pentose (ribose in RNA and deoxyribose in DNA), a phosphate group, and a nitrogenous base.

Oncogene. A gene that can transform normal cells into neoplastic cells; a gene capable of inducing cancer.

Oncovirus. Viruses that can cause cancer due to their ability to alter the cell cycle of the host cell.

p53 Protein: A nuclear phosphoprotein with a molecular mass of 53 kilodaltons (53 kDa), composed of 393 amino acids. It is a gene-regulating protein activated by DNA damage, aiding in the repair process by using checkpoints to halt cell cycle progression or induce apoptosis.

Phosphatases. Enzymes that catalyse the hydrolytic removal of phosphate groups from a molecule. In the cell cycle, these enzymes act to dephosphorylate CDKs and cyclin-CDK complexes, rendering them inactive.

Phosphatidylserine. A negatively charged phospholipid typically located on the inner layer of the plasma membrane's lipid bilayer. In apoptotic cells, however, it shifts to the outer layer, where it can serve as a marker, signalling neighbouring cells and macrophages to phagocytize the apoptotic cell.

Polypeptide. A linear polymer made of amino acids. Proteins are large polypeptides, and the two terms are sometimes used interchangeably.

Polysaccharide. A linear or branched polymer of monosaccharides; examples include cellulose, starch, glycogen, and hyaluronic acid.

Primary transcript: The immediate product of gene transcription, which will be subsequently processed to produce the final transcript(s).

Programmed cell death. A programmed "suicide" activated within animal cells, resulting in rapid cell death mediated by proteolytic enzymes known as caspases. It is related to homeostasis in the physiological regulation of tissue size. During this process, the cells shrink, the cell surface forms bubbles, and the cells break into membrane-bound fragments called apoptotic bodies.

Proteasome: A large protein complex in the cytosol with proteolytic activity, responsible for degrading proteins that have been marked for destruction through ubiquitination or other means.

Protein: The primary macromolecular component of cells, consisting of a linear polymer of amino acids linked by peptide bonds in a specific sequence.

Proto-oncogene: A normal cellular gene that, when altered by carcinogens, can gain the ability to function as an oncogene.

Pseudostratified epithelium. A lining tissue with a single layer of cells (making it a simple epithelium), but with cells of varying heights, resulting in the uneven distribution of their nuclei, giving the false appearance of stratification. Both tall and shorter cells rest on the basement membrane.

Repair mechanisms. Naturally occurring enzymatic systems in cells that correct errors in damaged DNA molecules.

Replicative cellular senescence: A phenomenon observed in cell cultures, where cell proliferation slows down and eventually stops irreversibly.

Reverse transcription: The transcription of RNA into DNA. In this case, the direction is opposite to that described by the central dogma.

Ribosomal RNA (rRNA): RNA molecules that form part of the ribosome's structure and participate in protein synthesis. Different types of rRNA are distinguished by their sedimentation coefficient, such as 5S rRNA, 16S rRNA, and 28S rRNA.

RNA polymerases: Enzymes capable of synthesizing an RNA copy from a DNA template and precursor nucleoside triphosphates.

Signal transduction: The conversion of a signal from one physical or chemical form to another. For example, the conversion of extracellular signals into intracellular ones.

Strain. A group of organisms developed from a common ancestor, sharing certain morphological, physiological, biochemical, or genetic characteristics (one or more of these traits) that distinguish them from other strains. Also referred to as a lineage.

Stratified squamous epithelium. A type of lining tissue made up of multiple cell layers. The basal cells are the deepest, resting on the basement membrane, and are spherical, cuboidal, or columnar, depending on the location of the epithelium in the body. The intermediate cells are elongated, while the superficial cells are flattened, resembling squamous cells. The name of this type of epithelium comes from the shape of the cells in the outermost layer.

Syndrome: A set of signs and symptoms that define the characteristic clinical manifestations of a disease.

Systemic: Referring to the entire organism as a whole.

Tautomerization: The rearrangement of electrons and protons within a molecule. In the case of DNA's nitrogenous bases, uncommon tautomers of adenine, cytosine, guanine, and thymine differ from their usual forms by the position of a hydrogen atom. The major issue with this alteration is that in their unusual forms, the bases can form incorrect pairings.

Teratomas: Congenital tumours (present before birth), made up of different tissue types. Teratomas in newborns are usually benign and do not spread throughout the body. They are often associated with a range of hereditary defects affecting the central nervous system, genitourinary tract, and lower spine.

TP53: A tumour suppressor gene located on the short arm of chromosome 17 (17p13.1), whose transcription product is the p53 protein. This gene spans 20 kb and consists of 11 exons, with the first being non-coding. Its transcription factor interacts with at least six other genes.

Transcription: The copying of a DNA strand into a complementary RNA sequence, mediated by the enzyme RNA polymerase.

Transcription factor. Any protein required to initiate or regulate transcription in eukaryotes. This includes gene regulatory proteins, general transcription factors, chromatin-remodelling complexes, and histone-modifying enzymes.

Transfer RNA (tRNA): Small RNA molecules used in protein synthesis as an adapter between mRNA and amino acids. Each type of tRNA molecule binds specifically to a particular amino acid.

Ubiquitin: A small protein found in all eukaryotic cells that degrades cyclins and other proteins, preventing cell cycle progression. Ubiquitin molecules covalently bind to lysine residues of other proteins. The attachment of a short chain of ubiquitins to a lysine residue marks the protein for intracellular proteolytic destruction by a proteasome.

Ubiquitin ligase: Any of the enzymes that attach ubiquitin to a protein, often marking it for destruction by a proteasome. The process catalysed by ubiquitin ligase is called ubiquitination.

Tumour viruses: Viruses that, once inside host cells, can make them cancerous.

Xenobiotics: Chemical compounds foreign to an organism or ecosystem. They can be found in places where they are not normally produced or expected. The term also applies to substances present at much higher concentrations than usual.

References

AIUB, C.A.F.; FELZENSZWALB, I. Os princípios do teste de Ames (*Salmonella* / microsomo) e sua aplicabilidade. **Genética na Escola** 06: 11-16, 2011.

ALBERTS, B.; JOHNSON, A.; LEWIS, J.; RAFF, M.; ROBERTS, K.; PETER, W. (2010) **Biologia Molecular da Célula**. 5 ed. Porto Alegre: Artmed, 2010, 1395 p.

AMABIS, J.M.; MARTHO, G.R. **Biologia volume 1: Biologia das Células**. São Paulo: Moderna, 2004.

AMABIS, J.M.; MARTHO, G.R. Biologia volume 3: **Biologia das Populações**. São Paulo: Moderna, 2004.

BALME, E.; THUILLIEZ, C.; LEJEUNE, T.; CHATEAU-ESCOFFIER, L.; BERNEX, F. Multiple atypical mucosal xanthomas in a dog similar to human verruciform xanthoma. **Journal of Veterinary Diagnostic Investigation** 21: 124–128, 2009.

BARIANI, M.H.; SOUZA, F.B.; TALON, D.B. Mastocitoma cutâneo em cães - relato de caso. **Revista Científica Eletrônica de Medicina Veterinária** 8, Janeiro de 2007.

BROWN, T.A. **Genética: um enfoque molecular.** 3 ed. Rio de Janeiro: Guanabara Koogan, 1999, 336 p.

CARDOSO, P.C.S.; ROCHA, C.A.M.; MOTA, T.C.; BAHIA, M.O.; CORREA, R.M.D.S.; GOMES, L.M.; ALCÂNTARA, D.D.F.A.; ARAÚJO, T.M.T.; MORAES, L.S.; BURBANO, R. *In vitro* assessment of cytotoxic, genotoxic and mutagenic effects of antimalarial drugs artemisinin and artemether in human lymphocytes. **Drug Chem Toxicol**. 42(6):608-614, 2019.

COSTA-LOTUFO, L.V.; MONTENEGRO, R.C.; ALVES, A.P.N.N.; MADEIRA, S.V.F.; PESSOA, C.; MORAES, M.E.A.; MORAES, M.O. A Contribuição dos Produtos Naturais como Fonte de Novos Fármacos Anticâncer: Estudos no Laboratório Nacional de Oncologia Experimental da Universidade Federal do Ceará. **Revista Virtual de Química** 2(1): 47-58, 2010.

COSTA-LOTUFO, L.V.; WILKE, D.V.; JIMENEZ, P.C.; EPIFÂNIO, R.A. Organismos marinhos como fonte de novos fármacos - histórico & perspectivas. **Química Nova** 32: 703-716, 2009.

CROSS, M.K.; POWERS, M.A. Learning about cancer from frogs: analysis of mitotic spindles in Xenopus egg extracts. **Disease Models & Mechanisms** 2: 541-547, 2009.

DA COSTA J.F.; LEAL, M.F.; SILVA, T.C.; ANDRADE JUNIOR, E.F.; REZENDE, A.P.; MUNIZ, J.A.; LACRETA JUNIOR, A.C.; ASSUMPÇÃO, P.P.; CALCAGNO, D.Q.; DEMACHKI, S.; RABENHORST, S.H.; SMITH, M.A.; BURBANO R.R. Experimental gastric carcinogenesis in *Cebus apella* nonhuman primates. **PLoS ONE** 6(7): 1-13, 2011.

DALLWIG, R.K.; WHITTINGTON, J.K.; TERIO, K.; BARGER, A. Cutaneous Mast Cell Tumor and Mastocytosis in a Black-masked Lovebird (*Agapornis personata*). **Journal of Avian Medicine and Surgery** 26(1): 29-35, 2012.

FARIAS, M.G.; CASTRO, S.M. Diagnóstico laboratorial das leucemias linfoides agudas. **Jornal Brasileiro de Patologia e Medicina Laboratorial** 40(2): 91-98, 2004.

FENICAL, W.; JENSEN, P.R. Developing a new resource for drug discovery: marine actinomycete bacteria. **Nature Chemical Biology** 2: 666–673, 2006.

FRANK-STROMBORG, M. "Cancer screening and early detection". In American Cancer Society. **A cancer source book for nurses**. Jones and Bartlett 4: 43-55, 1997.

GUIMARÃES, A.C. **Avaliação dos efeitos carcinogênicos e genotóxicos do urânio como indicador de risco à saúde de residentes dos municípios de Prainha, Monte Alegre e Alenquer no estado do Pará**. Belém, 2009. 81f. Tese (Doutorado em Neurociências e Biologia Celular) - Universidade Federal do Pará, 2009.

HANAHAN, D.; WEINBERG, R. The Hallmarks of Cancer. **Cell** 100: 57–70, 2000.

HANAHAN, D.; WEINBERG, R. Hallmarks of cancer: the next generation. **Cell** 144: 646–674, 2011.

HARVEY, R.A.; FERRIER, D.R. **Bioquímica Ilustrada**. 5 ed. Porto Alegre: Artmed, 2012, 520 p.

KOTERBAY, A.M.; MUTHUPALANI, S.; FOX, J.G.; MCNIEL, E.A. Risk and characteristics of gastric carcinoma in the chow chow dog. **The Canadian Veterinary Journal** 61(4):396-400, 2020.

LIMA, E.M.; RISSINO, J.D.; HARADA, M.L.; ASSUMPÇÃO, P.P.; DEMACHKI, S.; GUIMARÃES, A.C.; CASARTELLI, C.; SMITH, M.A.C.; BURBANO, R.R. Conventional cytogenetic characterization of a new cell line, ACP01, established from a primary human gastric tumor. **Brazilian Journal of Medical and Biological Research** 37(12): 1831-1838, 2004.

MENCK, C.F.M.; OLIVEIRA, M.C. Origem da vida: um tempo curto para uma experiência bem-sucedida. In: MATIOLI, S.R. (Ed.). **Biologia Molecular e Evolução**. Ribeirão Preto – SP: Holos Editora, pp. 11-14, 2001.

MEYERS, T.R.; HENDRICKS, J.D. Histopathology of four spontaneous neoplasms in three species of salmonid fishes. **Journal of Fish Deseases** 6(6): 481-499, 1983.

MOORE, P.F.; RODRIGUEZ-BERTOS, A.; KASS, P.H. Feline gastrointestinal lymphoma: mucosal architecture, immunophenotype, and molecular clonality. **Veterinary Pathology** 49(4):658-68, 2012.

MUKHERJEE, S. **O Imperador de todos os males: uma biografia do câncer**. Tradução: Berilo Vargas. São Paulo: Editora Companhia das Letras, 2012.

NUSSBAUM, R.L.; MCLNNES, R.R.; WILLARD, H.F. **Thompson & Thompson: Genética Médica**. Rio de Janeiro (RJ): Elsevier, 2008.

PEREIRA, R.C. Ecologia Química Marinha. In: PEREIRA, R.C.; SOARES-GOMES, A. (Org.) **Biologia Marinha.** Rio de Janeiro: Editora Interciência, p. 473-504, 2009.

RAMOS, P.; PELETEIRO, M.C. Três casos de neoplasias espontâneas em peixes. **Revista Portuguesa de Ciências Veterinárias** 98 (546): 77-80, 2003.

RAMOS, A.T.; NORTE, D.M.; ELIAS, F.; FERNANDES, C.G. Carcinoma de células escamosas em bovinos, ovinos e equinos: estudo de 50 casos no sul do Rio Grande do Sul. **Brazilian Journal of Veterinary Research Animal Science** [online] 44: 5-13, 2007.

RIBEIRO, H.F.; ALCANTARA, D.F.; MATOS, L.A.; SOUSA, J.M.; LEAL, M.F.; SMITH, M.A.; BURBANO, R.R.; BAHIA, M.O. Cytogenetic characterization and evaluation of c-MYC gene amplification in PG100, a new Brazilian gastric cancer cell line. **Brazilian Journal of Medical and Biological Research** 43(8): 717–721, 2010.

RIBEIRO, L.R.; SALVADORI, D.M.F.; MARQUES, E.K. (Org.). **Mutagênese Ambiental**. Canoas (RS): Editora da Ulbra, 2003.

ROCHA, C.A.M. **As pernas do caranguejo: *cancer crura*.** 112p. Belém (PA): [s.n.], 2013.

ROCHA, C.A.M. **Avaliação da genotoxicidade do cloreto de metilmercúrio em duas espécies de peixes neotropicais**. Belém, 2009. 90f. Tese (Doutorado em Neurociências e Biologia Celular) - Universidade Federal do Pará, 2009.

ROCHA, C.A.M. **Biodiversidade: genes – espécies – ecossistemas.** 172p. Belém (PA): Editora SEE, 2018.

ROCHA, C.A.M.; SILVA, F.P.E.; MARINHO, M.F.P.; LIMA, L.S.; BURBANO, R.M.R. Gene *MYC* e sua ativação no adenocarcinoma gástrico. **Genética na Escola** 19(2): 147-157, 2024.

SCHOTTENFELD, D.; BEEBE-DIMMER, J. Chronic inflammation: a common and important factor in the pathogenesis of neoplasia. **CA A Cancer Journal for Clinicians** 56(2): 69-83, 2006.

SCHUMACHER, J; BENNETT, R.A.; FOX, L.E.; DEEM, S.L.; NEUWIRTH, L.; FOX, J.H. Mast cell tumor in an eastern kingsnake (*Lampropeltis getulus getulus*). **Journal of Veterinary Diagnostic Investigation** 10: 101–104, 1998.

SNUSTAD, D.P.; SIMMONS, M.J. **Fundamentos da Genética.** 4ª ed. Rio de Janeiro: Guanabara Koogan, 2008.

SPORN, M.B. The war on cancer. **Lancet** 347: 1377–1381, 1996.

SZYF, M. The dynamic epigenome and its implications in toxicology. **Toxicological Sciences** 2007;100(1): 7-23.

THIRIOT-QUIÉVREUX, C.; WOLOWICZ, M. Chromosomal study of spatial variation of the prevalence of a gill neoplasia in *Macoma balthica* (L.) from the Gulf of Gdansk (Baltic Sea). **Ophelia** 54(1): 7581, 2001.

WEINBERG, R.A. **A Biologia do Câncer**. Tradução: Bruna Selbach ... [et al.]. Porto Alegre, RS: Editora Artmed, 2008.